# IN THE YEARS OF
# THE MOUNTAINS

# IN THE YEARS OF
# THE MOUNTAINS

EXPLORING THE WORLD'S HIGH
RANGES IN SEARCH OF THEIR CULTURE,
GEOLOGY, AND ECOLOGY

DAVID SCOTT GILLIGAN

THUNDER'S MOUTH PRESS
NEW YORK

IN THE YEARS OF THE MOUNTAINS:
*Exploring the World's High Ranges in Search of their Culture, Geology, and Ecology*

Published by
Thunder's Mouth Press
An Imprint of Avalon Publishing Group, Inc.
245 West 17th Street, 11th floor
New York, NY 10011

AVALON
publishing group incorporated

Copyright © 2006 by David Scott Gilligan

All photos courtesy of the author.

First printing, July 2006

Library of Congress Cataloging-in-Publication Data is available.

ISBN-10: 1-56025-836-5
ISBN-13: 978-1-56025-836-0

9 8 7 6 5 4 3 2 1

Book design by Maria E. Torres

Printed in the United States of America
Distributed by Publishers Group West

# Contents

# Maps

# *Introduction*

**THIS BOOK IS AN ODE** to the essence of high mountains. It is an attempt to investigate deeply the geography of high mountain landscapes in light of a thousand days on the rocks and snow, in the blistering sun, in the marbled clouds, and in the thin air. It is an attempt to braid the academics of geography with the observations and experience of exploration. It is about how we experience mountains, how mountains form, and how the elemental nature of rock, snow, ice, and air have come to define mountain landscapes. It is about how altitude simulates latitude, how mountains are sky islands in seas of lowlands, and about the origins and adaptations of alpine and arctic plants. It is about how people have coevolved with mountain landscapes over hundreds, even thousands of years, and how mountains, rather than cultural barriers, are actually cultural headwaters, where peoples of diverse languages, religions, and subsistence practices meet and mingle. But this book is also about wishing desperately for a good picket placement on a steep snowfield, a yawning crevasse just a few feet below. It is about watching a pious man offer burnt offerings of juniper to the god of the house while he hacks his lungs up after stooping all day indoors around a smoky dung fire. It is about being alone on a crystalline white summit during a temperature inversion, purple-gray clouds spreading out like an atmospheric ocean in all directions as far as the eye can see. It is about finding butchered caribou carcasses high on a glacier, the torn meat and hair surrounded by a hundred wolverine tracks.

There are four sections to this book, each of which focuses on a different mountain landscape. Each of these mountain landscapes represents a different continent and different attributes of the essence of mountains.

I begin in the Alps, because for millennia the Alps were the centerpiece around which Western culture twirled. The image of the Alps is burned so deeply into Western consciousness that the range has become the archetype against which all other mountains have been measured. Through explorations of the Bernese and Pennine Alps, I investigate the ancestral home of the wild, painted Celts and opportunistic Germanic tribes, the exploits of the intrepid romantic mountaineers of the Victorian age, the geologic history of the Alpine-Himalayan mountain system, and the glacial dynamics that have shaped the Alps of today.

I move on to the Himalaya, the biggest mountains in the world. Here I explore the Khumbu region and investigate the traditional cultural adaptations of the Sherpa people to high mountain environments, the simple but elegant tenets of the Buddhist tradition, human physiological adaptations to high altitude, and the impacts of high-profile Western mountaineering expeditions on the landscapes of traditional peoples. I also continue the theme of Alpine-Himalayan mountain building and investigate the physical evolution of the Himalaya.

Next, I travel far across the Pacific Ocean to the Southern Hemisphere and explore one of the most geographically isolated mountain ranges of the world, the Southern Alps of New Zealand. Here, the geographic factors that determine the distribution of plants and animals in mountain landscapes are exaggerated by the Southern Alps' geographical situation in the middle of the immense Southern Ocean. Through exploring the Arthur's Pass and Mount Aspiring regions, I investigate the biogeography of these small, distant "sky islands," how plants and animals arrived, adapted, and established themselves here, and the implications of the relatively late arrival of humans on the landscape.

Finally, I journey north along the North American Cordillera from subtropical latitudes to the subarctic. Along the way I explore the

Sierra Nevada of California, the North Cascades of Washington, and the Icefield Ranges of the Alaska-Yukon border country. I investigate the geologic history of the Cordillera, the longest and most extensive mountain system in the world. I explore the relationship between altitude and latitude, the similarities and differences between alpine and arctic environments, and the last holdout of the once vast glaciers of the Pleistocene. Moving north, I investigate the ideas of wilderness and wildness, as the landscape shifts from one where mountains are the last vestiges of wilderness, surrounded by and at times overwhelmed by civilization, to one where wilderness is the rule and civilization the isolated, vulnerable exception.

I stress that this is no armchair account based on maps, books, and scientific articles. In each section of the book I go directly to the land. There are stories of hard climbs, of clinking over steep, snow-covered rock faces with crampons on. There are stories of red-eyed ibexes, reflecting pools, and lonely walks in the rain. There are stories of high passes, yak hybrids, and meditation sessions at fourteen thousand feet. There are stories of crystalline pyramidal mountains rising out of cloudbanks off the Tasman Sea. There are stories of albino alpine wildflowers shaking in the dry wind. And there are stories of grizzly sows cocking their heads, cubs in the bushes just a few yards away. These images are reflections, and together they create a picture of one small man's experience of the essence of the high mountains of the world.

There are a thousand mountains to remember. Some of the ones I remember the most are the ones I never got to. Their stories remain shrouded in impenetrable mist and loom largest in my imagination. I find some solace in knowing that they will always be out there, that counting them is like counting stars, that there is always more to the world than we can hope to know, that even today we are still small, specks in a much larger picture.

# THE MEASURE
# OF ALL MOUNTAINS

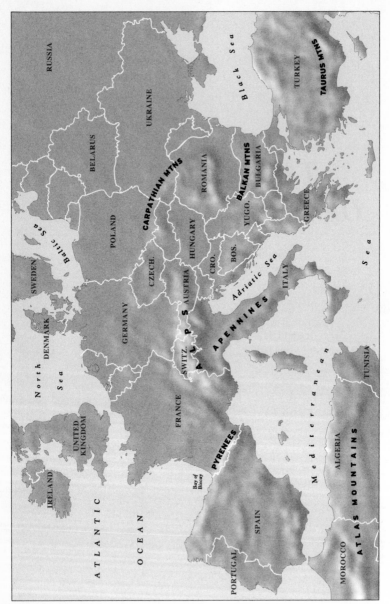

Europe

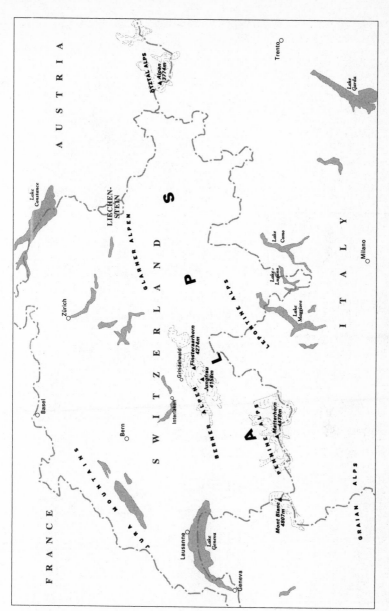

The Swiss Alps

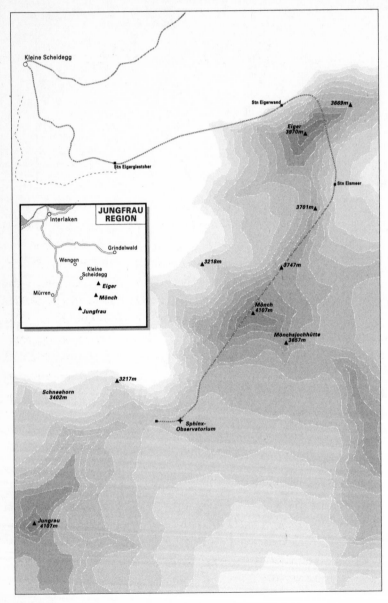

The Jungfrau region

# The Swiss Alps

## I

**THERE ARE TWO PILGRIMAGES** every mountain geographer must make. One is to the mighty Himalaya, the biggest mountain range on Earth. The other is to the majestic Alps, the archetype of the Western mind, the measure of all mountains. There are dozens, perhaps scores of mountain ranges in the world that are compared to the Alps. In fact, alpine areas across the globe are so named because of their close affinity with the Alps. The word *alpine* derives from the Latin for *white*, used by the Romans to refer to the snow-covered treeless peaks that bound their realm, and later used to describe the inhabitants of the region. Eventually the word found wider use, and has since been used to describe similar snow-covered mountains all over the world. But comparisons are comparisons. There are no substitutes. A pilgrimage to the Alps means to prostrate oneself at the feet of those mountains against which all others, even the Himalaya, even the Andes, have been measured—the place of neck-craning vertical relief, glistening white snows, cracked blue glaciers, fertile green valleys, gushing meltwater streams, frozen Icemen, blue-painted Celts, bronzed Romans, German-speaking, French-speaking, Italian-speaking, Romansch-speaking peoples, edelweiss, fondue, lederhosen, tinkling cowbells, watches, cheeses, bank accounts, and romantics. There was never a question of whether or not to go. Only a question of when.

\* \* \*

I awake from a comalike slumber after a thirty-six-hour, six-thousand-mile day, crawl laboriously out of the blind of my tent, and lie down on the cool, damp grass of the Gletscherdorf campground. In front of me stands the most imposing, vertical, dramatic mountain front I have ever seen. Cumulus clouds roll right into the wet, gray limestone of the north-facing wall of the mountains, green-splashed here and there with tenacious vegetation, and roll up beyond the last line of trees to the glacier-covered peaks above, swirling, marble gray in the shade and bright in the sun, adding immeasurably to the drama of the scene. The great wall is split, and out of it pours a shimmering valley glacier that seems to grope for the front of the mountains, where in later years it might have found relief from its confinement, and spilled out all over this place in broad lobes clear to the Swiss Mittelland. The only sound is that of the swollen, pale gray and white meltwater stream just yards from me. It makes no diversions, no bends, and heads straight downslope, a low-gradient flume. It is as young as these mountains, and as vigorous. I can't help but make comparisons, and familiar mountain ranges come to mind. The Front Range of Colorado. The east side of the Tetons. The North Cascades. The eastern escarpment of the Sierra Nevada. The Southern Alps. The Scottish Highlands. The mountains of western Norway. But the Alps before me eclipse them all.

I have never been here before, but already I know these mountains. The glacier is the Oberer Grindelwaldgletscher. The peaks are the Fiescherhorn, Klein Fiescherhorn, Eiger, Mönch, and Jungfrau. For nearly three decades, they have made appearances in my life in picture books, and later textbooks, and now they become real. The distance of pictures dissolves, and as it does I reconcile the fiction of distance with the reality that is in front of me. I am a romantic and more often than not I find myself disappointed when confronted with the reality of a place I have mused about in my imagination for a long time. But that is not the case today. Instead, I find myself silent with humility, searching for words to describe my reaction to the scene before me, but unable to find them.

Mike Spayd has been awake for awhile and has already made a trip to the Nescafé machine at the campground store. He sees me up and out, comes over with his weird, purposeful walk and good ol' boy smile, and says, "Damn, Dave, this place is *un*believable." He looks like a slender version of Emilio Estevez. I can't think of anyone I'd rather be here with. Mike is one of the most energetic, motivated, dedicated people I know, and our relationship with mountains is very similar. His Alabama fraternity-boy background gives him a sense of fun and loyalty that I admire. Mike was a student of mine for two years and since then we've worked together, climbed together, skied together, and spent well over a hundred nights in the mountains in each other's company. I trust him as I trust few people.

We meet the streets of Grindelwald after cups of black tea with milk and sugar, which I need just to sift through the haze of the last day's traveling. We poke around the resort town in search of information on getting up high. Grindelwald is like a Swiss version of Stowe or Aspen or Mammoth Lakes. By "Swiss" I mean it is older, cleaner, smaller, and aesthetically perfect. The tourists dabble in curio shops full of Swiss army knives, cowbells, red T-shirts with white crosses, and Norwegian-style sweaters; bakeries stocked with sweet pastries and delicate breads; restaurants advertising fondue at atrocious prices; and even a Mexican-style discothèque. Small, neat cars buzz by in small, tidy lanes and everything is amazingly efficient and hopelessly attentive to detail, the result of thousands of years of smart thinking. Grandiose wooden houses, chalets, and hotels line the lanes and dot the nearby green hills and fields, all proud and square with massive overhanging roofs, heavy Germanic ornate woodwork gracing the eaves, brightly painted wooden shutters framing the windows, impossibly perfect bouquets of geraniums bursting forth from wooden flower boxes in splashes of pink, magenta, fuchsia, and scarlet. The architecture complements the tweed coats and vests, pipes, knee-high socks, lace-bottom britches and leather hiking boots of the traditional European hikers, who are mostly of the generation passing, and it contrasts sharply with the conspicuously large and modern building in the center of town, which bears a sign in equally conspicuous English that reads THE SUNDANCE HOTEL.

Soon we realize the dark side of traveling in Switzerland. While looking at a guidebook in a bookstore, we are confronted by the nervous shop owner; she makes it clear we are not to look at books if we don't intend to buy them. We've had the cover cracked for all of two minutes. I figure a nation so tight and efficient must have rules, and Switzerland in particular must have a lot of them. And most people actually follow them. From quiet hours to where you put your feet on the train to looking at mountaineering routes in overpriced guidebooks, the Swiss have it sewn up, and in a single day I am being schooled on all these matters. Tourism was invented here, and the antiquity of the establishment has resulted in a kind of cultural ascendancy among the Swiss who inhabit villages such as Grindelwald. It is almost as if it is your privilege to be able to give them your francs. Dependency on tourists' money has inevitably caused some resentment among the locals. It is a classic love-hate relationship, and with few dollars to spend, I find myself far from the "love" end of the spectrum. We close the book and return it to the proprietor, having written down three words: *Jungfrau from west*. She grips the book with a sneering, sarcastic remark and takes it with her behind the desk. We leave the shop with a mixture of feelings, shamed by our apparent poverty and lack of etiquette, but glad to know we will have our introduction to the high Alps tomorrow.

In less than twenty-four hours, we are over ten thousand feet above sea level, traversing the glacial flats away from Jungfraujoch, the highest railroad station in Europe. We are already a little bit giddy after the steep, circuitous train ride up the Bernese mountain front, through the wall of the Eiger, and up to the Bernese divide. We have never in our lives done anything but walk up into the high mountains, typically laden with heavy packs full of equipment and supplies for a week or more, and the idea of any kind of mechanized support leaves us wondering if we are cheating. I ponder the ways of the backcountry of North America, where the best mountains are the ones that take three or more days to get to, and the time spent out in the field is reason enough to be there. But the Alps have long been known for a completely different ethic. Here, camping anywhere below the permanent snow line is forbidden and camping above the snow is scorned. Faced with the choice of spending the night in the overcrowded Mönch Hut, its whereabouts indicated by the road-wide trail of packed snow and steady stream of people headed towards it, or returning to the valley for the night, we choose the latter.

The grandeur of the Alps from below was stunning, but this is something else. We are aghast. I am wide-eyed and thinking of the Antarctic or the Himalaya. We are surrounded by immense, dark, jagged, youthful mountains mantled with snow, and glaciers of a scale I have not seen before. The Cascades or the Southern Alps suggest this level of mountain glaciation but do not approach the reality of the Bernese Alps. Cirque glaciers are pasted to mountain walls at angles I didn't think were possible. All of them, scores visible from the west ridge of Jungfrau, are tributary to an immense valley glacier, striped by an accumulation of medial moraines, that winds its way south and out of view. There are many such glaciers in the Alps of Switzerland, home to eight of the ten largest glaciers in the larger Alpine chain. But even among them this glacier deserves special respect. It is the Grosser Aletschgletscher, the largest and most extensive glacier in the Alps, the headwaters of the Rhône River. If we were to follow the glacier down to its snout, negotiate the churning meltwater as it issued forth from beneath the ablating ice, and somehow follow the rushing river downstream, we would find our way down the Valais and into France, down the fertile Rhône Valley to the ancient trading center of Marseille, where we would spill into the briny waters of the sparkling Mediterranean.

We gain the east rib of the Jungfrau massif via several sections of tricky, steep snow, but a trail is in, so we are able to negotiate the terrain with relative ease, without crampons, without ropes, using only our ice axes and kicking steps through a shallow veneer of new snow to the hard permanent snow beneath. The snow sections are

separated by a short band of rock, and upon reaching it we are glad
to have our feet unencumbered by points. We scramble up to more
snow, kick steps, set shafts, and rise up to a low-angle dome. At its
head is a final steep slope that leads up to a saddle, a window to the
east, and there is a brazen blue sky over it. I rest a short while and
watch as Mike ascends behind me, a small thing in a landscape vast,
white, gray, and bright.

Just below the saddle we follow a broken trail through a gap in an
otherwise enormous cornice, a ten-foot standing wave of snow with a
perfect curl, accentuated by the sugarlike coating of yesterday's new
snow. To our right the cornice dwindles, but only where we climb
through is it broken, and it rises up with the terrain on either side, left
to a curvaceous snow-dome, right to the southeast ridge of Jungfrau.
We pull ourselves up to the windowsill for a look out on the world.

The mountains drop steeply to the east, as if they have been pulled
out from under us, down, down, down to the deep green of the
Lauterbrunnen Valley nearly twelve thousand feet below. Crowning
the valley are numerous peaks, and from their flanks white glaciers
issue forth, flowing down and into the valleys, becoming gray in their
descent, even blackened with debris, cracked and broken, failing.
From out of each glacier a roiling meltwater stream bursts forth,
quickly tumbling downslope and free-falling in horsetail cascades over
hanging valley lips, streaming down in a dozen milky strands, joining
into one brilliant, white river below, and disappearing into the distant
dream of the fertile valley. I can't believe this place is real, but it is. Mike
and I sit for a long while and say nothing. Before us, in all directions,
lies the geography of the Alps, which, taken far enough in all directions, is the geography of Europe.

The Alps are an immense mountain range spanning over 600
miles in length and 150 in width. Like most of the ranges that bol-
ster the underbelly of Eurasia, they trend east-west, creating a huge
mountain arc from the Mediterranean coast near Nice, France, to
deep into eastern Europe. As a mountain chain they are composed
of many dissected subranges, or regions, the delineations between

which there is often discrepancy, but seldom in defining the three highest, most glaciated, and magnificent regions: the Mount Blanc region of France, the Pennine Alps of the Swiss-Italian border, and the Bernese Alps of the north slope. These latter two regions form the heart of the Alps.

The Alps are the centerpiece of Europe both physically and culturally. Physically, they form headwaters and thus the divide between all of the major watersheds of continental Europe. The Rhône, the Rhine, the Danube, the Loire, the Seine, and the Po all rise in the Alps, finding their uppermost sources in the glaciers and snowfields of the high mountains. During Pleistocene glacial advances, these upper river valleys were inundated with and subsumed by glacial ice that accumulated to depths in excess of a mile and poured out across the lowlands of south-central Europe, covering nearly all the ground of Switzerland and much of the surrounding land area.

European cultures also find their headwaters here. From time immemorial, the Alps, like all mountain ranges, have formed a geographic boundary between cultures. But a boundary is also a meeting place on either side of which cultures evolve in relative isolation from one another, developing rich and diverse language and lifeways, meeting at times on the high passes to exchange goods and ideas. In the dim past the Alps were the divide between Stone Age cultures, spread still farther apart by the advancing Alpine icecap, brought closer together during times of retreat. In historical times, the Alps divided Celtic Europe to the north from their more orderly Roman neighbors to the south. Today, seven European countries meet in the Alps, including France, Switzerland, Italy, Liechtenstein, Germany, Austria, and Slovenia. In Switzerland alone, the four different languages spoken, German, French, Italian, and Romansch, reflect the cultural diversity found in mountainous regions, and reinforce the idea that mountains, most often thought of as cultural divides, have long been cultural meeting places.

Among the Alps, the Bernese stand out as being the coldest, wettest, most heavily glaciated region of the greater chain. Their northern front is as steep an escarpment as any in the world, dropping thousands of feet down in a handful of miles. It is crowned by the massifs and associated ridges of the Jungfrau, Mönch, and Eiger, peaks as infamous as any in the world, and dozens more whose names remain obscure to most people. This wall forms a formidable barrier to prevailing weather systems coming in from the North Atlantic and bears the full brunt of those storms, even exacerbating them through orographic lift to the extent that precipitation levels on the northern mountain front may be as much as ten times that of the nearby lowlands, and as much as eight times more than others areas of the Alps, such as the Valais to the south. All of this precipitation, combined with the cold temperatures typical of high mountains at forty-seven degrees north latitude, makes for the relatively heavy glaciation of the Bernese Alps. But although the ingredients for the ice come from the north, the topography of the region dictates that the longest and most voluminous glaciers all flow south of the northern front, where snows collect and accumulate in the lee of the high peaks and ridges.

The Bernese Alps also stand out as being one of the only regions of the Alps that is not crossed by international boundaries. The whole of it lies within Switzerland. But not even the diversity of Switzerland reaches far into the region, and the Bernese Alps are almost entirely Germanic. Only in the extreme southeast of Bernese country is the French influence easily recognized. But it was not always so. Passes still bearing the stones of ancient Roman roads cross the divide in some places, remnants of cultural exchange in times long past.

We put our crampons on and prepare to go. Facing the southeast ridge of Jungfrau, its elusive summit obscured by our very nearness to it, we begin the final ascent, following the ridge up a gentle curve that soon turns to steep snow. Wet metamorphic rocks are exposed to our left, where the mountain drops most precipitously, and we periodically step onto them, observing metal stakes with prefabricated anchor rings, the likes of which I have never seen. We pass them by, stopping to breathe frequently in the thinning alpine air. But the ascent is straightforward, and we continue, unroped. Eventually the angle lessons and the very summit of Jungfrau comes into view. We walk the final steps together, right up to the very edge of the summit, the Bernese Alps laid out like a living map before us. But even as we arrive, clouds begin to billow up and obscure the vividness of our surroundings, swirling up to the high peaks as we watch, transfixed. Earlier parties having abandoned the summit hours before, we make no haste to leave. The air seems electrified, cool, thinner, more precious. In thirty minutes the nearby Monch is enveloped in a thin, wispy shroud, a lenticular cloud obscuring only its uppermost summit. Moments later the Eiger is cloaked. We look around us one last time and begin our descent, elated at our introduction to the magnificent Alps.

*left*, Grindelwald and the Grindelwaldgletscher
*right*, Rottalsattel

## II

The weather forecast is monotonous, calling for rain today, rain tomorrow, and rain the next day. This means snow up high for the next few days, which translates to potential avalanche conditions in the high mountains. With an already existent desire to explore the habitable regions of the Alpine countryside, we decide to take this opportunity to spend some time with wildflowers, tinkle-bell'd cows, farmhouses, and rural people. We set out in the rain, leaving Grindelwald a second time for a few days' walk across the Bernese Oberland.

Somehow we take a wrong turn at a train depot called Alpiglen, and instead of ascending promptly to the high route that follows the very base of the Eiger, we are funneled into a double-wide tourist track headed straight for the next train depot at Klein Scheidegg. Large groups of Japanese tourists that seem to move as units swarm around us. Almost every one of them is moving downhill as we move up, and it strikes me that this must be some kind of classic, widely published day walk: get off the train at Klein Scheidegg, walk down the entire way to Alpiglen, or perhaps even Grindelwald. No sweat, no strain. Each time we pass through a group there is a strange feeling that I can only relate to passing through the midst of a large group of migrating ungulates. No eye contact is made. There is no communication. As we approach each other, the crowd splits and passes closely on either side of us, then rejoins just below us, continuing downslope as if nothing had ever happened. It is a strange feeling, that of being a casual obstacle in the path of a collective, psychologically distant consciousness. This happens half a dozen times, and each time is much the same. By the time we reach the crowds at Klein Sheidegg, we are bewildered, and quickly look for a way out.

Finally we are in open, green country, a mosaic of conifer forest and alpine meadow, and the tourist crowds of a mile ago are quickly forgotten. The forests are dark, composed mostly of a tight canopy of Norway spruce with splashes of red-fruited rowan here and there, and occasional stands of larch. Where the conifers open up even a little bit, an array of heaths form an understory, and I am reminded of a dozen places I have been before. Heaths do well in the acidic conditions of the conifer forest floor, where a constant supply of the chemical *lignin*, found in the conifer needles, is released into the organic layer of soil through decomposition. In most places in the Northern Hemisphere, where there are conifers, there are also associated heaths. Here the association is between Norway spruce and thick-leaved evergreen heaths. In Scotland it is between Caledonian pine and heathers. In Maine it is between spruce-fir forest, white pine, and blueberry. In Arizona it is between ponderosa pine and man-zanita. In the Sierra Nevada it is between lodgepole and western white pines and red mountain heather. In the Yukon it is between white spruce and bearberry. The list goes on. Everywhere in the forests of the North the association is the same, but the players are different and specific to place.

In the rain, the forest seems oppressive, though it offers some respite in the form of shelter. When we emerge into the open of alpine meadows we are hit by the downpour. It looks and feels like Scotland on a bad day, and I am left wondering how places like Scotland and Norway have developed reputations for cold, wet weather and Switzerland has managed to dodge such accusations. Perhaps they carefully and intentionally omit such facts from the outgoing lit-erature in some effort to protect the tourist industry. Then I remember where I am: the Bernese Oberland. The clouds shift and a hole opens up to our left. Above the rolling green of the alpine meadows, a sheer wall of gray limestone becomes visible. Pouring down its face is a steep, reckless hanging glacier. As if by some gift of providence, a lone shaft of sunlight penetrates the gloom and illumi-nates the face of the ice to the point of brilliance. I have to squint to look at it. This is not Scotland.

As we weave in and out of treeline we pass near several low-lying alpine farmhouses. They are consistently well built, made of wood with some stone, with low-angle roofs. I speculate that the intention of their relative prostrate form is to keep their exposure to weather minimal. But I wonder about those roofs. They don't look like they would shed snow so well, which makes me wonder if wind conditions up here are such that snow accumulation is not a big issue. But if snow doesn't accumulate here, then where does it? And what about drifts? I am left thinking the buildings must be so solidly built that they can just hold their snowload. They are seldom occupied in the winter months, originally designed as summer shieling houses for shepherds and cowherds to accompany their flocks to the high country while the lowland valleys were cut for hay. A few have signs on them advertising tea and dairy products for the passerby, and linens hung out to dry in the cold rain.

There is livestock everywhere. When we can't see them we can hear them, bells tinkling in the mist, and I am reminded that those heavy bells, each and every one poured and stamped by hand, are not just for looks. I try to imagine how many hundreds of thousands of times those bells have helped lonely cowherds find their mist-shrouded cattle on days just like this one. When the mist clears, I see bovines munching away at the greenery, or lying down quietly in groups, while views of hanging waterfalls leaping down limestone walls open and close around them. I am reminded of the open rangeland of Nevada, where gaunt cattle chew scrub and occasionally cactus in the blistering Great Basin sun, and it strikes me that Swiss cattle have it made. The resiliency of this land, due to a combination of high rainfall and resulting high productivity, is amazing. That, combined with the small scale of the livestock operations, means livestock grazing, even in high alpine meadows, has the potential to be sustainable and can even promote, rather than decimate, diversity among plant species.

I am spotted by a score of goats who head straight for me in antic-ipation of I don't know what. I hold out my hand and they take turns sucking on it.

Even though we came expecting to see this, it still comes as a bit of a shock. As North Americans, we almost unconsciously associate being in alpine country with being in wilderness, and the idea of being in the midst of cow herds and farmhouses evokes a strange mix of emotions, including nostalgia and frustration. But Switzerland is not the anomaly, North America is. North America is, in fact, the only habitable continent that has major high mountain ranges in which people do *not* live. Everywhere else, from the Alps to the Himalaya to the Andes, people have been a part of the mountain landscape for hundreds, if not thousands, of years, and developed coevolutionary relationships with mountain landscapes over long periods of time. Here in the Alps, humans have never been a "visitor who does not remain," but rather an integral and longstanding citizen of the landscape.

When human ancestors first set foot on European soil, the Alps and the surrounding lands were periodically inundated by the vast Alpine icecap. At this time in the long story of human evolution, *Homo erectus*, the first hominid to leave the confines of Africa and spread across the subtropical underside of Asia, had evolved into archaic *Homo sapiens*, complete with enlarged, rounded skulls that housed brains the size of a modern human's, and specialized voice boxes that allowed for complex vocal communication. But the *Homo sapiens* line to which we belong had two lines in its early stages of development, two subspecies that show up as distinct in the fossil record. One was *Homo sapiens neanderthalensis*, the other was to become *Homo sapiens sapiens*. The Neanderthals specialized in colder climates more than did their slender cousins, and dominated much of Europe and parts of western Asia. Their robust, stocky physique meant a lower surface-to-volume ratio, and this adaptation made them more heat efficient. Despite popular belief, their brains were actually larger than ours, particularly in the area of the cerebellum. Equipped with sophisticated stone tools, the Neanderthals spread north as far as northern Germany and Kiev. They formed complex social groups, laughed and prayed, buried their dead, and lived much like their eventual successors, the Cro-Magnon *Homo sapiens*.

However widespread the Neanderthals may have been from two hundred thousand to forty thousand years ago, it is not known if they ever actually inhabited the Alps. The earliest evidence for human occupation of the region comes from the Alpstein massif of north-eastern Switzerland. Here, limestone caves at six thousand feet above sea level yield the remains of a Paleolithic hunter-gatherer culture that spent time there a hundred thousand years ago, during a pronounced warming period. But cooling temperatures and advancing glaciers soon chased them out, and humans did not return for many thousands of years. It may be speculated that earlier incursions occurred, and that humans were more widespread than the isolated find in the Alpstein suggests, but if that is the case, subsequent glacial advances have since wiped out all traces. Meanwhile, the Neanderthals were eventually subsumed by the more technologically advanced Cro-Magnon people, who wielded compound tools and had even more specialized voice boxes. It is not clear whether the newcomers decimated Neanderthal populations or interbred and eventually absorbed them, but the latter implies that some mix of Neanderthal and Cro-Magnon blood is at the root of European genetics.

As the last vestiges of the Pleistocene Alpine icecap withdrew, slurries of silt-laden meltwater washed out the steep-walled, flat-bottomed Alpine valleys of today, depositing nutrient-rich soils into these troughs. Warming continued, punctuated by brief periods of relatively cool temperatures, from fourteen thousand to six thousand years ago, when global climatic conditions reached a marked warm point known to climatologists as the Climatic Optimum. Throughout this time, glaciers receded and advanced, but the overall trend was recession, ultimately revealing the vast and previously uninhabited mountain landscape of the Alps. At the height of the Climatic Optimum, the amount of ice mantling the Alps was significantly less than today, and many of today's Alpine glaciers did not exist.

As the glaciers pulled back, Mesolithic hunter-gatherers followed the newly exposed soil, seeking unclaimed hunting grounds. These people likely came from the Po River plain in north-central Italy (to the south), and Southern Germany (to the north), from which they followed herds of migrating ibex, chamois, and even reindeer. At first they had only stone tools, and hunted and gathered, but as time went on they adopted the metallurgy practices of adjacent areas, and acquired domesticated animals such as sheep and goats. Eventually they learned to cultivate cereal grains. They cleared and burned forests, planted crops, tended animals, and settled in. By the time of the Climatic Optimum, Europe had become increasingly agrarian, and was spattered with small-scale agricultural settlements in most of the major river valleys.

At this time, four to six thousand years ago, it is believed that humans made their first serious inroads into the high country of the Alps. This seems late at first, considering that the continent had been inhabited for over 150,000 years, but the landscape of the Alps is colder and more rugged, and therefore much less accessible to primitive peoples. Across the globe, both alpine and arctic regions were the last to be inhabited by humans, even when adjacent land was occupied for millennia. In North America, the continent was inhabited for ten thousand years before humans developed the technologies necessary for life in the high Arctic, and Native Americans never established permanent habitations in the high mountains. In lower Asia, one of the oldest seats of human culture, people did not permanently occupy the alpine regions of the Himalaya until the last millennium. The prerequisites for even temporary camps or excursions into such environments were sewing awls, for making weatherproof garments, and compound tools for firemaking, hunting, fishing, toolmaking, and tailoring. To actually live year-round in such environments, highly specialized tools, (including weapons), clothing, and traveling technology were required, and in later cases innovative pastoral and agricultural practices.

The 1991 discovery of a 5,300-year-old corpse in the Italian-Austrian Alps shed new light on the technological advances necessary for humans to inhabit the high Alps. The Iceman was a media sensation and an anthropological revelation, not because of his antiquity—remains thousands of years old are not uncommon—but rather because of how intact everything associated with the find was. Among the artifacts found in association with the Iceman were a copper ax (still bound to its handle by leather straps), a yew longbow (still in the process of being made), a quiver full of arrows (some of which still had fletching attached), a leather belt-pack including several flint points and knives, as well as tinder, a birchbark container filled with grass and maple leaves, grass-lined leather boots similar to Inuit kamiks or Sámi finnesko, and the remains of fur leggings, a fur coat or cloak, a bearskin hat, and a grass overcoat. Everything the Iceman needed came from the landscape he lived in: bone, wood, fur, plant, feathers, copper, and stone.

Equipped with such an array of specialized tools and clothing, by then anthropologically fully modern humans began pasturing their flocks in alpine meadows, hunting in the conifer forests that darkened the flanks of the mountains, and utilizing high passes as trade routes between tribes in adjacent valleys. And so the story of cultural geography in the high Alps began.

Bernese Oberland

# III

We pass through more patches of forest and meadow, heading ever east across the wide valley of the Trümmelbach. Ahead, the land gives way, and at first we are confused until we realize our valley is hanging, and the Lauterbrunnen, into which this valley feeds, is an abrupt three thousand feet below. We come to the upper lip of the Lauterbrunnen, and the land drops. It is well beyond a precipitous drop. It is vertical. Before us and below us is the archetypal steep-walled, flat-bottomed glacial valley. Its sides are of ice-hewn limestone. Waterfalls, innumerable in this rain, line the far wall like white, gushing veins streaming down the blackened cliffs to the rampant river below. It is a limestone version of Yosemite, but even more perfectly U-shaped. But below, there are no park curio shops or visitor shuttle busses. There are instead houses, driveways, and stubby European cars, a community of people whose past in this valley reaches back to centuries before Europeans even knew there was an America. And above, seemingly pasted to the side of the mountains just atop the vertical rock faces, are the twin villages of Mürren and Gimmelwald. They likely began as seasonal farmhouses, the summering grounds for the Germanic people who spent their winters in the valley bottoms. It has been centuries, but the traditional seasonal lifestyle still exists. Threatened with extinction by a modern industrial culture that renders traditional farmers economically inefficient and redundant, the Swiss government provides subsidies to its rural communities to maintain the cultural landscape and keep tradition alive.

I am struck by this picture and can't help but juxtapose this Alpine valley with its granitic counterpart in North America. Looking in this valley, I am reminded of a place in my country where you have to pay to get in, can't take as much as a leaf away for fear of getting a ticket from a pistol-bearing law enforcement ranger, are regulated as to where you can and can't sleep, or even walk, and are so alienated from the landscape that you can't take or make anything from it and are but a visitor who does not remain. Here, I look down on people's homes,

a valley that has been tilled and hayed for perhaps millennia, meadows
that have been grazed as long, and forests that have been harvested as
long. There is no gate, no entrance fee, and I can walk anywhere I
want (but can't sleep anywhere but indoors) without bumping into or
being interrogated by an armed official. The people I see are an integral part of this landscape and have been since time immemorial. But
every square acre is used, accounted for. The major river valleys are
managed for industry and agriculture. The Alpine valleys are managed
for agriculture and winter fodder. The forests are managed for timber
and wood products. The Alpine meadows are managed for grazing.
Even the glaciated mountains are managed, for recreation. There is
nothing that comes even close to a blank spot on the map.

The story of human presence in the Alps is a long one. From early
on, people made incursions into the region but seldom stayed more
than a fortnight or perhaps a season. But as competition for resources
in the adjacent lowlands increased, cultures were pushed up against
the mountains, and eventually, largely in order to avoid competition,
they figured out how to make a permanent living up high. The story,
like most good stories, is both convoluted and fascinating.

The Iceman of the 1991 archeological find lived at the close of the
Neolithic, during a brief period of human prehistory known as the
Copper Age, or Eneolithic. That he lived in or near the Alps is an
important point, because at that time most of the easily discovered
deposits of copper occurred on exposed mountainsides. The Iceman,
then, may well have been at the cutting edge of metallurgy technology. Eventually, the know-how spread, and soon after humans figured out how to smelt copper they were making their first alloys,
combining copper and tin to make bronze. This important innovation heralded the Bronze Age, often associated with the classical
period of Greek history that inspired the *Iliad* and the *Odyssey*. With
the rise of metallurgy, the forests of Europe were exploited for timber
to keep the smelters burning, and the result was often strife and warfare between developing civilizations. Control of timber resources
and trade routes became increasingly important, and competition for

these often resulted in migration or warfare. But bronze was a soft metal. Swords bent and broke, shields and greaves caved in, spear points quickly became dull. Eventually, people figured out how to make charcoal, and this enabled them to smelt an even harder metal. That metal was iron.

The progenitor of both warfare and migration always begins with a technological advance. With new technology, resources can be exploited more efficiently, and two things can happen. First, surplus foodstuffs result in a population increase. Second, as in the case of metallurgy, more efficient exploitation quickly leads to resource depletion. In both cases, the result is stress and the solution is either to migrate and find new resources, thereby avoiding competition with neighboring tribes, or to compete directly, which means war. Most often in history the reality has been some combination of the two. That a culture has never voluntarily lessened its dependence on a resource (which might seem like the simplest solution) is a question to be reckoned with.

Two technological innovations characterized the Iron Age and heralded a slough of massive human migrations (and associated warfare) across Europe. One was the development of iron implements, and the other was the widespread domestication and use of the horse. The former required wood fuel and the latter required pasturage. Those cultures that mastered iron and the horse and controlled the resources necessary to sustain these technologies ruled Europe. Two such dominant cultures existed in the latter centuries BCE. To the north were the wild, heroic, rural, decentralized Celts. To the south were the well-ordered, disciplined, highly centralized Romans. Between them were the Alps.

Both the Celts and the Romans had their origins in the Indo-European peoples of eastern Europe, the Russian steppe country, and west-central Asia. As these ancient peoples became increasingly successful, they displaced neighboring cultures with astonishing rapidity in a series of massive migrations that have seldom been repeated in history. Over the centuries spanning the second and third millennia

BCE, the Indo-Europeans split into divergent groups, and language and culture also diverged. Some groups migrated south and east, eventually passing into the Indian subcontinent. Other groups went south and west, into the Mediterranean basin. Still other groups went due west, into western Europe. By the time the Celts reached Europe and the Roman republic was formed, these two peoples spoke mutually unintelligible languages and had few similarities in customs. But both had iron, and both had horses. And Europe gave way before them like grass before a fire.

From around 750 BCE, the Celts dominated all the land from the Alps north to the North Sea (and eventually Britain and Ireland), and west to the Atlantic, pushing indigenous cultures of these areas, such as the Basques and Albans, to the outermost fringes of the continent, destroying or absorbing whoever didn't give way before them. But however unified they were by language and custom, the Celts were never unified politically. They are characterized by history as being a fierce people, agrarian but known for spontaneous, massive migrations, a warrior culture fond of single combat between champions, ritualized psychological warfare, and heroic displays of bravery. But the Celts were also a deeply spiritual people with a complex mythos, steeped in poetry and song, a sophisticated oral tradition, and some social norms regarding property and gender that would put even the most progressive societies of today to shame. By 500 BCE, Celtic tribes inhabited the major valleys in and around the Alps, and began making incursions into northern Italy.

In the south, Roman culture was just budding, largely at the expense of the Etruscans to the north. In contrast to the Celts, the Romans were highly centralized, bound by political doctrine, and above all, orderly. Though it would be centuries before Rome became an empire, already it was making a reputation as a city-state to be reckoned with.

In was only a matter of time until these two cultures met. Eventually, backed up against the Alps, it was the Celts, the older and more extensive of the two cultures, who spilled over the long-standing mountain barrier and into the sphere of influence of the

Romans. In the fifth century BCE, Celtic migrants began moving into the Po River valley of northern Italy. They came by the thousands. Though at first the migration was a peaceful one, soon the Celts' population grew and, sandwiched between the Alps and the Etruscans, began taking lands to the south. The Etruscans fell easily before the long swords of the Celts, and the dwindling civilization soon found itself applying for help from its longtime enemy to the south. Rome responded by sending envoys north to parley with the Celts. At first, negotiations went well. The Celts were impressed by the Romans, who seemed an important state to be reckoned with. Because the Romans sent envoys and not armies, the Celts accepted their offer of peace, provided the Celts could take land in the region of the Etruscans. If the Romans refused, the Celts would initiate an attack immediately so the Romans could witness firsthand their indisputable superiority. When the Roman envoys asked by what right the Celts took land from the Etruscans by force, the response of the Celts was plain and simple: "To the brave belong all things."

All might have gone without bloodshed had the Etruscans yielded before the will of the Celts, but they did not. When opposed, the Celts took up arms. Contrary to the international law of the times, the Roman envoys joined in the fight, and in the fray one of the Romans used his influence to get close to a Celtic chieftain, and killed him.

The Celtic reaction was swift and decisive. They spiked their hair with lime paste, painted their faces, and headed for Rome. Thousands of Celts traveled eighty miles in four days for the purpose of avenging their chieftain. The spectacle of these men and women on the road to Rome was astonishing. Blond, fair, painted, and naked, singing in unison and yelling passionately, the Celts overwhelmed the region, and the Latins, (Latin-speaking peoples surrounding Rome who at that time may or may not have been considered actual Roman citizens, or "Romans") largely undefended, shrank before them. But the Celts were not interested in outlying villages. They marched straight to Rome without stopping. Eleven miles from the city they met the first and only Roman defense, and surprised the Romans by brilliantly attacking the core of the Roman forces, driving them back into supporting ranks, and scattering the entire army. The Romans, mortified by the Celtic war cry and overwhelmed by their tactics, fled in terror.

When the Celtic leader Brennus arrived with his people at the gates of Rome, they found the place empty. Suspecting a trap, they sat tight and waited. They couldn't yet imagine that the entire city had been evacuated and the people relocated in the more defensible Capitol. After three days, Brennus realized the place had been abandoned, and he and his army walked into a hauntingly silent Rome. There, one of the most macabre moments in history awaited them. In an abandoned yard, perched high on elegant ivory chairs, sat a group of old, bearded men. They sat motionless, scepters in hand, the senior patricians of the city-state of Rome. The Celts thought them idols, and walked in a circumference around them, awed and bewildered. But eventually a warrior reached up and plucked a hair from one of the statues, and the statue came to life! The now animated patrician arose and struck the Celt on the head. Astonished, the Celt drove his sword through the patrician, and within seconds the surreal scene was washed in red.

The Celts lay siege to the Capitol for seven months, and eventually demanded a thousand pounds of gold as ransom for Rome. The Romans scraped their coffers clean to try to meet the demands of the Celts, and when unable to, they accused the Celts of using false weights to counter the balance. At this, the Celtic chieftain Brennus threw his heavy longsword into the tipped scale, adding its weight to that of the Roman gold. As his iron clanged against the scale he uttered, "*Vae victus*," "Woe to the vanquished."

Rome never recovered from its initial interactions with the Celts, and for centuries the Romans looked to the Alps and their high passes with fear. And their fear was well justified. In 217 BCE, 173 years after the Celts sacked Rome, the Carthaginian leader Hannibal fell on Italy from the French Alps and ravaged the countryside. At his side were thousands of singing, painted Celts.

Eventually, as Roman influence over Italy waxed, the Celts were pushed out of the Po River valley and into the Alps, from which they waged decades of guerrilla warfare on their longtime enemies. But the effect of centuries of living (and dying) with the Celts was that the Romans developed the most extensive, well-organized army that had yet been known to the Western world, and even after hundreds of years, the humiliation the Celts had wrought on the early Roman republic was not forgotten. Meanwhile, to the north and east, migrating Germanic tribes began to encroach on Celtic territory. As Rome pushed north from Spain and Italy, and the Germanic tribes pushed south and west, the Celts of the continent were caught between and forced to consolidate into what is now Switzerland, France, and Belgium, then known to the Romans as Gaul. Where once the Alpine passes had served as trade routes between tribes and military inroads into Roman territory, now the Romans gained control of them, built roads, and so crossed the great cultural divide and gained access to inner Europe.

By the middle of the first century BCE, the tables were turned completely. Julius Caesar, then an aspiring politician and aggressive military general, was given the responsibility of the Celtic-inhabited areas surrounding the Alps. In 58 BCE, seeking advancement and renown, Caesar attacked the Helvetti Celts, who were in the process of a massive migration from the Alpine valleys of what is now Switzerland to the open land of adjacent Gaul. Caesar's armies attacked and killed over six thousand men, women, and children, killed all of their livestock, burned the hundreds of wagons the Helvetti had made in preparation for migration, and destroyed the two years' worth of crops they had accumulated in anticipation of the move. The survivors made a weary trek back to their lands in Switzerland, but there was little there for them. The Celts themselves had burned their own homesteads upon leaving so they would not be tempted to return.

In the end, Rome destroyed the Celts utterly, and by 50 BCE, the culture that had once dominated Europe was either pushed off the Continent to Britain and Ireland or was absorbed into the Roman world. The Romans graded and paved the alpine passes, bridged the rivers of the Alpine valleys, and drew all of western Europe into their sphere of influence. Romanized, the surviving Celts of the Continent lost all of their cultural autonomy, and most of their identity, including their language. They became the French, who to this day recognize Vercingetorix, the last chieftain to unite the Celts against Caesar, as their first national hero.

But it was the Germanic tribes that had the last word. Defeated time and time again by the Roman legions, new waves of Germanic peoples seemed to flood continuously into Europe from the east. Early on, these peoples had made inroads into the Alps, pushing Celtic tribes out or warring with them over resources. When Celtic Gaul fell to Rome, Germanic mercenaries fought on the side of Caesar. But eventually, the Germanic tribes turned their swords on their liege lords, and as the Western Roman Empire waned due to a combination of political strife, depleted resources, and weakened legions, the Goths and Visigoths, as well as the Huns and Vandals, ransacked Roman Europe. Rome fell, and Europe was up for grabs.

In the wake of the fall of Rome, continuing waves of Germanic migrants moved into what is today Switzerland. In the west, where the last vestiges of Romanized Celtic culture still held on, the new-comers were quickly assimilated, but in the east, where Celtic influence had not been strong for centuries, the Germanic peoples dominated. The Germanic tribes, like the Celts before them, were cattle herders, and employed an array of agricultural and pastoral practices for subsistence. They quickly found the advantages of utilizing Alpine meadows for summer grazing, thereby freeing up the valley bottoms for more intensive agriculture. By 800 CE, tribes originating from the Bernese Oberland crossed over the Grimselpass and the Lotschenpass. By draining upland bogs and clearing forest, they were able to establish the first permanent settlements above five thousand feet in the Alps. Continuing innovations drove cowherds to the very snout of glaciers in search of pasturage, and thereby freed up increasingly more of the valleys for agriculture or haying.

For several centuries, these Germanic peoples spread out over the Alps, until by the end of the medieval period all of the high Alpine valleys had been brought under the teeth of German dairy cattle. The social structure of these early tribes, as was the case in Scandinavia at the time, dictated that the family lands be passed on to a single heir. This meant that all other siblings had to marry into an existing family or seek out new lands on their own. In Scandinavia, this led, in part, to the Viking migrations. In the Alps, it led to the Walser migrations. The Walsers, whose name comes from *Walliser* (German for *Valaisian*—the Valais was the valley where they were centered), were favored by feudal landowners due to their innovative land management practices, and were given privileges for opening up previously unclaimed territory. Thus, the Germanic peoples made still deeper inroads into the once Celtic- and Latin-speaking Alps, and eventually came to dominate the region. As a legacy of these migrations, Switzerland today is over 60 percent German-speaking.

When the Swiss Confederation was formed in 1291 as a resistance to the Austria-based Habsburg dynasty, other regions rallied to join the cause. With the prospect of being dominated by foreign overlords becoming common in late medieval Europe, the idea of a confederation of supporting communities seemed utopian. In less than a century, the number of communities, called *cantons*, that together comprise Switzerland had more than doubled, and they continued to increase until 1515, when, confronted with the prospect of curtailing the autonomy of its cantons in order to support a more unified expansionist policy, the Swiss Confederation decided to withdraw completely from the international scene. They renounced expansionist policies and declared neutrality. With the Alps as their centerpiece, they absorbed the influences of every culture around them, facing and often defying the more dominant nations of France, Germany, Austria, and Italy. It was radical. From the beginning, the confederation bore the Latin name *Confederatio Helvetica*. The language is Roman, but the name is that of the Helvetti, who were Celts.

We descend steeply, down towards the Yosemite-like valley of the Lauterbrunnen, and eventually we are drawn into the gorge of the Trümmelbach. Here, the rushing water has cut deeply into the water-soluble limestone bedrock and formed a narrow cleft, a slot in the valley wall only six feet wide, but a hundred feet deep. The water is barely visible in the cloudy evening light and seems to disappear into the mysterious incision it has sculpted. It roars, turbulent in its confines. A fine mist surrounds the gorge, augmenting the incessant rain. We descend still further, and the rocks become so steep that soon we are grabbing at old, fixed cables to negotiate the wet, exposed, smoothed limestone.

We emerge onto the flat, green valley floor just next to where the Trummelbach surges out of its hiding place in a flood of milky water. It hits the valley bottom and gushes straight out to the Weisse Luts-chine, the master stream at the valley's center. We walk across the green grass, through the backyard of someone's house, and onto the slick, wet road. A small car zips by, heading downvalley, and its occupant seems to take no notice of our soaked, weary condition. We are in need of shelter from a day of mountain storms. The people of the valley sit comfortably in their homes, just as they have for thousands of years.

Lauterbrunnen Valley

# IV

Zermatt is like a carnival or a shopping mall mixed with an old historic village. Old stone streets are lined with crowded shops full of tourist trinkets. In between are more tolerable tea houses and sweet-smelling bakeries, separated by old, massive wooden buildings that date back to the sixteenth and seventeenth centuries. Behind the main street front, still older buildings remain hidden from the mainstream crowds, old houses and granaries no longer in use. These buildings are invariably propped up on pillars, and slate-stone discs a meter wide separate the pillars from the wood of the house. The discs served as rodent protection for centuries simply because rodents would have to crawl upside down and around them to gain access to the building, and rodents cannot readily do that. But the windows are boarded up, and though these buildings have stood for centuries through rain, snow, wind, and fire, the inhabitants of Zermatt have traded them for more modern accommodations.

It has rained or snowed every day since our arrival in Switzerland. Here, in the Pennine Alps, the annual precipitation is a fraction of what it is in the Bernese Alps—less than a fourth, to be precise. Friends we made in Grindelwald assured us that the sun would be shining on the Matterhorn and blue skies would greet us when we arrived. Instead, the sky is deep gray, and the light at midday feels like evening. Somewhere out there the mountains are steadily accumulating snow, and avalanches are loosening. But for now that doesn't bother me, and my thoughts are confined to the rain at hand. Scotland and Norway are beach towns compared to this. My tent has begun to leak like an ill-fitted boat turned upside-down. September is reputed to be the best month for stable weather in the Alps, and thus far the only stability has been that it has rained just about the whole time. So far, more rain has fallen here in a few days than falls in Arizona in a year. I believe the glaciers must be growing.

We walk through town, and then back toward the train station in search of the local campground. We begin to make light of our situation, realizing the hopelessness of complaining. "We should've made this a rafting trip!" Mike suggests. Feeling comically cynical, I wonder about the whereabouts of the elusive campground. "Watch, we'll probably end up in a vacant city lot somewhere, like that one." I point to the right. "Or that one." I point ahead. And there are tents, dozens of them, soaked by the rain, in the middle of a soggy city vacant lot. "Camping Matterhorn," the campground we have been looking for, is a slug-infested mudpit. We set up our leaky tent and stow our belongings inside, leave the premises, and head back into town.

If Zermatt seems like the archetypal tourist town it's because it is. Tourism was largely invented in the Alps, and even among Alpine villages Zermatt gets quintessential status. The disadvantages are crowds, prices, lack of local flavor, tourist-wary proprietors, and that all-too-familiar feeling of being rutted into some prefabricated, inauthentic, mindless trap. But there are advantages, too. The mountains loom so large around the village that they are omnipresent, and a short walk out of town in almost any direction brings you right to their feet. And the place is so steeped in mountaineering lore that it has become an inseparable part of the cultural identity. Catholics go to the Vatican to see the hub of their culture. Mountaineers go to Zermatt. The reasons reach all the way back to the Renaissance.

After centuries of medieval introversion, the reintroduction of cosmopolitan ideas to a conservative Europe wrought tremendous changes in the worldview of the populace. But while nations such as Spain, Italy, Portugal, France, and England consolidated and vied for power through exploration and exploitation of trade and land resources in the Indies and the New World, Switzerland consolidated and dug in at home, taking in bits of the surrounding countries here and there and eventually stopping expansion altogether. Particularly in those Swiss lands in and around the Alps, the life remained a simple agrarian one, little different from the way it had been for hundreds of years. Isolated from one another by high, often snowbound mountain passes, valleys developed their own dialects, styles, and tastes. Although all a part of Switzerland in name, the varied regions of the *Confederatio Helvetica* had closer cultural affinities to the countries they faced than to the cantons on the other sides of the passes. Northern and eastern Switzerland remained Germanic. Western Switzerland remained French. Southern Switzerland remained largely Italian. And a small portion of the southeast held on to its Romansch traditions.

As the European powers waxed tremendous wealth amassed both on the Continent and on the British Isles. Rationalism and Newtonian mechanization brought the clock, the pulley, and eventually the gears of industry. Asian nations that had long been the envy of European artisans and merchants alike began turning to their European neighbors to gain expertise in growing areas such as mechanics and engineering. This combination of surplus wealth (largely due to the unabashed exploitation of furs, mineral wealth, and slaves, most of which came from faraway continents), and developing expertise in mechanical and engineering fields (afforded no doubt as a result of peasant-class and slave labor), contributed to a perhaps unforeseen and certainly unprecedented cultural development in Europe. That development was the Industrial Revolution.

Agriculture, and eventually extractive practices such as mining, in combination with trade goods, provided the raw material on which the machines of the Industrial Revolution fed. As such, where agriculture had historically been driven by subsistence, it became increasingly driven by business. For the first time in many an Alpine farmer's memory, prices rose and fell, following the ever-fluctuating whims of the market. Instead of laying in harvest goods for winter, farmers found themselves bringing their goods to market a more profitable practice. Cash crops replaced subsistence crops and farmers quickly felt the bite and sting of the erratic market economy. Where climatic fluctuations once dealt alternating hands of affluence and poverty, now market fluctuations dealt those same hands with equal ambivalence as to the consequences. Perhaps most important, more than ever before, the value of raw goods paled in comparison to that of hard currency, and rural economies that once had a relatively high degree of autonomy were brought into the fold of a cash-based economy. The feudal system that had long been emplaced on the Continent had to evolve. Lords turned into businessmen, and the fruits of their tenants' labor a precious commodity. But the rift between classes became less and less distinct as the effects of industrialization reverberated through the years, and the old forms of social hierarchies were about to be challenged.

As an increasing amount of natural resources were imported from faraway places and the market continued its fluctuations, human populations in Europe underwent a series of radical changes, minor migrations that resulted in major demographic shifts across both the continent and the British Isles. As industrial innovations strained resources on the home front, feudal landowners evicted tenants in exchange for sheep pasturage to support the burgeoning wool industry. Boatloads of Scottish Highlanders were shipped off to the New World on old, rotten ships, many of which foundered on the open sea and never made the far shore. Other rural folk were displaced to urban areas closer to home, and sought work in factories. Still more left their farms by choice, seeking the alleged benefits of urban life. Mechanized tools displaced hand tools, and an ever-increasing array of labor-saving devices contributed to radical changes in the very nature of work. Despite often squalid living conditions in the expanding urban centers, the progress of industry continued, and eventually Europe saw a new class of people emerging. A middle class.

For untold centuries, it had always been the rich, and only the rich, who could afford the pursuit of leisure. One of the remarkable aspects of the Industrial Revolution was that for the first time since the Roman Empire, and perhaps since the glory days of the original leisure cultures, the hunter-gatherers of the Paleolithic, large numbers of people were afforded opportunities to pursue activities other than work. It did not happen overnight, nor did it happen everywhere. It was a gradual transition, although certainly abrupt in the grand scheme of things. And it happened first and foremost in those countries that were most industrialized and ahead of the economic curve. Among such countries, the first and foremost was England.

But the progress of industry did not go unchecked, and by the late eighteenth century, amid the cries of praise for industrialization, there emerged cries of complaint. In response to the emergent rational-mechanical worldview, what began as a few small voices speaking out from the grind and crank of the factory grew into the public outcry of multitudes. Leading the way of this new movement was the Genevan philosopher Jean-Jacques Rousseau, who eloquently argued for the value of human feelings and moral sense, both of which the rationalism of the Enlightenment threatened to eclipse. Included in Rousseau's doctrine was an aesthetic and sentimental approach to nature and all things natural, illustrated by the philosopher's inspired accounts of mystical union, and further articulated in his numerous philosophical writings. The seeds of discontent must have hung ripe in the air. Soon others caught on. In America, the transcendentalists took up their pens. Ralph Waldo Emerson, in his book *Nature*, went so far as to espouse the notion that nature was a reflection of God. Henry David Thoreau furthered this idea by suggesting that nature might even *be* God. Thoreau also took Rousseau's original idea of the "noble savage" several orders of magnitude further when he asserted, in his 1862 essay "Walking," that "in wildness is the preservation of the world." The Romantic movement, contesting in all ways the pure rationalism and mechanization of the Industrial Revolution, was in full swing.

As the nineteenth century waxed and reached its noontide, an ever-increasing number of English men and women, equipped with the new notion of vacation and afforded the financial opportunity to pursue this notion, and espousing a mainstream version of Romantic ideals, headed for what was left of Europe's "natural" places in search of spiritual and aesthetic sustenance. They followed in the footsteps of such renowned Alpine travelers as Jean-Jacques Rousseau, Horace Benedict Saussure, Wolfgang von Goethe, Lord Byron, William Turner, Charles Dickens, and Mark Twain. While nature philosophers of the Continent had initiated the rediscovery of the Alps as the ultimate Romantic landscape, it was primarily British notables, in a flurry of written publications and landscape art, that brought the

Alps into vogue and opened the mountains to a steady flood of
tourists out of Victorian England. Artists, writers, aspiring philosophers, trendy vacationers, and eventually mountaineers flocked to the
stupendous rock battlements, cascading waterfalls, and unearthly glaciers of the Alps. The Swiss, in the heart of it, turned their barns into
inns and opened their purses. And the money came rolling in.

Alpine tourism was brought to its mythological zenith during the
golden age of Alpine mountaineering, which spanned the decade
1855–1865. During this relatively short time, many of the most
prominent peaks amongst the Alps were climbed for the first time by
pioneering British mountaineers. But in nearly all cases, the Englishmen
enlisted the aid of local Swiss mountain guides, the sons of the shepherds of the Alpine valleys, born of families who had lived in the Alps
for hundreds of years. But however long these rural people had lived
in the shadows of the highest peaks of the Alps, none of them had
ever climbed to their uttermost heights. Such intrepid adventures
were left to the restless Victorian English.

The British began the development of Alpine climbing techniques
back home in the Lake District and the Highlands of Scotland, and
wasted no time in transferring and perfecting these techniques in the
higher, snowier, icier, more vertical Alps. Dressed in wool and oiled
canvas, and equipped with hobnail leather boots, wood-shafted ice
axes, and hemp ropes, the Englishmen challenged the rock walls, glaciers, snow chutes, and corniced ridges of the high mountains. Their
methods were unrefined, yet for the most part they sufficed. In the
place of the pitons, pickets, and ice screws of today, they relied on
courage, grit, and luck.

Dufourspitze of the Monte Rosa massif, the highest peak in
Switzerland and second only to Mont Blanc in all of western Europe,
was climbed by the Englishmen Grenville and Christopher Smyth in
the summer of 1855. In the years that followed, the English journalist
Sir Leslie Stephen made several ascents of peaks in both the Bernese
and Pennine Alps. By the 1860s the British were biting into more
ambitious projects. The mighty Weisshorn went to the stalwart John

Tyndall, who later turned his attention to other mountains elsewhere in the Alps. Indeed, at the height of that decade it seemed as if no mountain could stand in the way of the resolute Englishmen. Their names are forever attached to those peaks they climbed. Interestingly, little if any attention is given to the local Germanic and French herders who guided them along their way.

Not everyone was so transfixed by the allure of mountaineering. It was, after all, largely a recreational pursuit, the Alpine region having long been known by its inhabitants. It lacked the depth of exploration that concurrent expeditions deep into Africa and the polar regions had, through which European minds awakened for the first time to entire continents and ocean basins, and cultures unknown or long forgotten. To some, mountaineering was mere sport, the restless exploits of wealthy urbanites who had strayed so far from nature. Charles Dickens eschewed the activity, commenting that mountaineering "contributes as much to the advancement of science as would a club of young gentlemen who should undertake to bestride all the cathedral spires of the United Kingdom." Dickens may have had a point, but he was no mountaineer and had little firsthand experience on which to base his scathing critique. The image of British schoolboys painfully straddling cathedrals is a far cry from the first ascent of the Matterhorn.

On July 13, 1865, Edward Whymper, accompanied by seven others, including Lord Francis Douglas and two mountain guides, set out from Zermatt to attempt an ascent of the as yet unclimbed Matterhorn. The elusive peak had become an icon among mountain icons, a symbol, rivaling if not overshadowing Mont Blanc, of the magnificence and ruggedness of the Alps. Whymper had attempted its precipitous flanks before and failed, tried again, and still failed. His summer objectives met, Whymper turned his attention back toward what he likely considered some unfinished business, collected his climbing party and his wits, and made his way upward.

Whymper and his party made the base of the mountain that day and made camp at eleven thousand feet. His guides, Peter and Croz, scouted the route, and on their return reported little difficulty, saying,

"We could have gone to the summit and returned today easily!" Whymper must have been gripped by anxiety. An Italian party had set out from Breuil, Italy, two days before, and the prospect of seeing backlit figures on the summit loomed large in Whymper's mind. They laid in and spent the night, shivering as usual in the thin wool wraps that were the sleeping bags of the times. Whymper and Lord Douglas slept inside the tent, while the others slept out in the open mountain air. In the morning the party shook off the chill of the night and set out for the peak, sending the younger son of one of the guides back to Zermatt before beginning the ascent.

The party climbed quickly, following the same route the guides had scouted the day before, reaching fourteen thousand feet before 10.00 A.M. Continuing up the east face, they followed the route of least resistance, eventually traversing to the north slope, where the real climbing began. Here, Croz and Whymper led, with the less experienced men following, and the guide Peter in the rear. They ascended a moderate snow slope, clinging to ice- and snow-encrusted rocky handholds. After some zigzag route finding, they finally rounded a corner and climbed the last two hundred feet of snow to the summit. Whymper and Croz, unroped, literally raced to the top. Once on its uttermost height, they quickly scanned the surrounding snow. There was no longer any doubt. No human being had ever set foot on the Matterhorn before.

Elated at their success, giddy in the thin air, Whymper and Croz, upon spotting the Italian party on a ridge over a thousand feet below, began shouting and hooting, trying to get the attention of the Italians. The leader of the Italian party was known and respected by Whymper, who later wrote, "He was the man, of all those who attempted the ascent of the Matterhorn, who most deserved to be the first upon his summit. He was the first to doubt its inaccessibility, and he was the only man who persisted in believing that its ascent would be accomplished. It was the aim of his life to make the ascent from the side of Italy, for the honour of his native valley." But Whymper was there first, the Italians mere specks on the ridge below, and the Englishman never hesitated to take the honor of the first ascent, however deep his respect for the Italian may have been. As emotions gained lift and began to soar, Whymper began trundling rocks downslope, exhorting Croz to do the same. Eventually both men found themselves prying rocks loose with their ice ax shafts. "A torrent of stones poured down the cliffs," Whymper wrote. "There was no mistake about it this time. The Italians turned and fled."

As the others arrived, Croz, lacking a summit flag, tore off his own blouse and tied it to the tent pole he had planted on the highest point of the mountain. It was seen far below in Zermatt, and also in Breuil, and both towns thought their man the first to stand atop the Matterhorn. The men who were on top built a cairn, then stopped moving awhile, and the view opened up before them. One hundred thirty miles in all directions, the crown Europe sprawled out before them, a vast ocean of snow-clad mountains.

The party began their descent in midafternoon. They roped up for the difficult section of downclimbing below the summit ridge, but for some reason they neglected to anchor their manila rope on the rock, as they had originally intended. Whymper brought up the rear of the party, at first tied only to the younger Peter, but as the ascent continued he tied in to the larger group. They ascended slowly, carefully, moving one man at a time over the precarious terrain. Croz was in the lead, guiding the Englishmen Hadow, Hudson, and Lord Douglas. The elder Peter, his son, and Whymper followed. Croz laid his ax aside to lend aid to the less able Hadow. As Croz turned his attention back to the mountain, the unthinkable happened. Hadow slipped and took Croz with him. Then Hudson was pulled off. Then Lord Douglas. The elder Peter, his son, and Whymper held fast, and for a moment it may have seemed that they had stayed the fall. But the rope broke, and the four men fell. Whymper wrote, "For a few seconds we saw our unfortunate companions sliding downwards on their backs, and spreading out their hands, endeavoring to save themselves. They passed from our sight uninjured, disappeared one by one, and fell from precipice to precipice on to the Matterhorn Glacier below, a distance of nearly 4,000 feet in height. From the moment the rope broke it was impossible to help them."

Within days, all of Europe had heard of the disaster on the Matterhorn. Whatever lens the collective consciousness of the West had been using to view the world was broken, and whatever holy light the European mind had cast on the high mountains went dark. A shadow lay across the Alps, the specter of death that overnight took the luster off of the Golden Age. But were those reckless men who wasted their lives among the glistening snow-scoured heights, or were they heroes? Would Europe learn from the example of the Matterhorn disaster, or would the possibility of a noble death in the mountains make the pursuit of mountaineering even more alluring than before?

We walk through town, rain drizzling down onto dark brown wooden buildings heavy with age, pooling around the street stones before finding its way into culverts and ditches. Wandering, we pass by the local cemetery. We stop, and the sounds of the village seem to fade to a hushing rain, voices distant and obscure. It is an eerie place, and the deep, ashen gray of the sky only enhances this mood. Between the rails, in a small patch of wet green grass and rain-slickened stones, lie the bones of scores of climbers who gave their lives to mountains like the Matterhorn, the Weisshorn, and the Monte Rosa. But the numbers of people who have died in the Alps are far more than can be counted here and would fill this cemetery a hundredfold. Each year for a hundred years more graves are dug and filled, and in recent decades the number of climbing dead each year is greater than the number of days of summer. It is a sobering thought. Today, tomorrow, or the next day, someone—maybe a dozen people—will die in the Alps.

In the Alpine museum, we seek shelter from the rain and sift our way through rooms full of old wooden skis and long wood-shafted axes, hobnail boots, yellowing photographs, and dusty old accounts of famous ascents. There is a sense of intrepid adventure in the eyes of men who have long since passed away, and occasionally a hint of other emotions, like fear, and perhaps even regret. In the case of Whymper, who, following the Matterhorn disaster, put down his ice ax and lived to grow old, there is regret. On the wall of the Matterhorn room, cased in glass, are his words, written one month after his attainment of the summit:

> *People talk about the vanity of earthly wishes and we have all felt at some time or another that they are vanity, but never have I felt it as I do at the present. For five years I have dreamt of the Matterhorn; I have spent much labor and time upon it— and I have done it. And now the very name of it is hateful to me. I am tempted to curse the hour I first saw it; congratulations on its achievement are bitterness and ashes and that which I hoped would yield pleasure produces alone the severest pain; it is a sermon I can never forget.*

Around me, people in wet raincoats scuffle across the floor noisily. I sit low, against the wall, and scratch Whymper's words in the pages of my journal. At first I am agitated at the human traffic that seems so busy around me. I grow frustrated and want to cry out. But something rises in my throat and I try to hold it there. My chin quivers, and a single teardrop hits the ink-stained page of my journal.

Downtown Zermatt

# V

Looking out from a rock perch on Pfulroe, across the long, stretched, cracked-in-a-thousand-places, debris-flecked surface of the Findelgletscher, itself flanked by high-stacked lateral moraines where the tops turn to soil over thousands of years and green things tenaciously hold root, a herd of chamois graze on nectar-filled, late-blooming alpine flowers and succulent green grasses. I have been chasing them from above in hopes of a better view, but they move far more deftly over the rocky terrain than I do, and they seem to be getting further and further away. Above them, talus slopes tinker and loosen, sending fist-sized chunks of gray stone tumbling from the rockbound shores of tiny, green tarns to the more luxurious plant-spattered terrace below. Still higher, the clouds tear open to reveal the glimmering forms of the Monte Rosa massif, the Adlerhorn, and the Strahlhorn, opposite the cloud-wreathed yet mysteriously visible archetypal form of the infamous (and rightly so!) Matterhorn. Far away from cable cars, gondolas, huts, restaurants, and trains, bivouacked in a sheltered col, I recall with lucidity why I came to the mighty Alps of Europe, and I see at last that vision unveiled and revealed before me. After days of rain and snow the sun has returned to grace the mountain landscape with its golden autumnal light and illuminate its fluted recesses before our cloud-weary eyes, and we revel in it like fogbound sailors at their first sight of land.

We are perched at the edge of the nival zone, that realm of ice and snow where few organisms survive other than lichen and glacier fleas. Here, new snow feeds the starving glaciers even in summer, and the stark, sterile landscape bears its white wintry cloak the year round. From out of the snow and ice rise exposed peaks and ridges, where snow perpetually blows or is sloughed off the bold rock. Here are the high mountains, the lofty peaks, the crown jewels of the Alps.

Last week's weather has covered the crevasses and loaded the mountain slopes with a foot of fresh, powdery snow. The chances of punching into a crevasse or getting caught unawares in an avalanche have multiplied tenfold. Many of the high peaks, including the omnipresent Matterhorn across the valley, are in full winter dress and will not be climbed again until next summer. The Swiss mountain guides are hanging up their ropes, and many are saying that winter has already arrived. Whether or not this is true, travel is slowed tremendously by the surficial snow, and most of the rock sections of the mountaineering routes have become too dangerous to safely ascend. Yesterday, traversing on the Italian side of the divide toward the twin peaks of Castor and Pollux, we were met by two British climbers making a hasty retreat. They had triggered a slide and it rightfully spooked them off the mountain. On the return, one of them had twice punched into a crevasse. We took the hint, and found some satisfaction on a relatively mild and safe walk up the Breithorn via its west ridge. It seemed all too likely that the Alps would be shut down for the season, but we weren't ready to go just yet.

We set out this morning on foot from Blauherd, just above the treeline, in hopes of initiating a multiday project involving the noble yet relatively obscure Rimpfischorn. Contrary to Alpine style and etiquette, we carried everything with us that we would need to sustain ourselves for three days in the high mountains. By European standards this was something of an expedition, and those people we passed looked at our hulking packs with wonder. But we were just doing things the way we were used to, carrying our food, shelter, clothing, and equipment, living and sleeping outside in the high mountains, having wide expanses of landscape to ourselves, and loving life because of it. The fact that camping below the permanent snowline was illegal only made it more appealing. The idea of wilderness camping in the Alps was breakthrough. We had to do it.

We made camp at the col between Spitzi Flue and Pfulroe peak, on a flat gravel bed tucked away in the rocks just below the permanent snow line. Here we enjoyed a lunch of baguette with brie and tomato, which after the twelfth day in a row was just as good as the first time. We spent the afternoon scouting a route across the Langfluegletscher, waiting for the clouds to part so we could see the rock towers of the

Rimpfischhorn. But its secrets remained hidden from us, and eventually we shrank away and back down to our small camp, where we made hot tea with milk and sugar and huddled among the wet rocks.

On Pfulroe, alpine crows roosting on the rocks below rustle about noisily. There is a break in the weather and the view of the surrounding mountains is precious. We have not consulted any route descriptions, and in hopes of a more traditional approach to climbing the mountain we consult the land instead. The map suggests following the west ridge, staying on the upper Langfluegletscher, climbing up and over a series of rock bands, and accessing the upper massif of the Rimpfischhorn via the west face. The land seems to confirm this idea, but the upper massif is still wreathed in lingering clouds. I doubt myself. The cold of evening sets in. I wander down the clinking rocks in search of warmth.

Six thirty A.M. I awake with a cold nose inside the sugar-coated tent. In the night, the rapidly cooling air had released one last gasp of precipitation and sheathed the tent in ice. As the last clouds, relieved, let fly into the night, the temperature dropped, and even in my half-sleep I knew what had happened out there, and that when I awoke again the sky would be crisp and cold and bells would be ringing.

I emerge from the tent singing songs, much to Mike's dismay, who throws rocks at me from the red bundle of his bivvy sack not far away. The air is fresh and winter cold, the sky a stark and transparent deep blue. But even now, dark clouds brood heavily over Italy and send wisps over the Pennine crest. We move slowly but deliberately, eating hot oats and packing our rucksacks for the day's climb. Soon we are walking, up and over the col, and stepping onto hard snow.

As we cross the Langfluejoch, the snow proves excellent for walking, and the terrain is moderate enough that neither crampons nor ice axes are necessary. The snow surface glimmers in the morning light, the well-formed stellar ice crystals reflecting the welcome rays of the sun like a million delicate stars set down in brilliant white, sweeping spans. But we seek the shade that keeps the snow firm, knowing the return will likely be in ankle-deep slush.

We ascend a series of low-angle rock outcrops amid the snow-covered glacial ice, and as I step across the slick, frosted surface of terra firma I watch sequences of long-since metamorphosed sea-floor sediments pass beneath me, once part of the mighty Tethys Sea that separated an older version of Europe from an older version of Africa. Moving swiftly, I take mental notes. Silvery, soapy, flakey schists border serpentines and old, crunched pillow lavas laden with pea-sized garnet crystals. I can't help but pick up a few samples. There are octahedral magnetite crystals, black and hard, polished and perfect. Mixed in are elongate green mineral crystals, presumably hornblendes, most of which are connected to paintbrushlike splays of asbestos crystals. I am enamored, but I keep moving. I am always amazed at how much of the Earth's surface is covered by old sea floors. And here, the old sea floors bear evidence of intense and tortuous agitation and uplift.

Leaving the sandwiched, uplifted, ice-polished ancient sea floor behind, we venture back out onto the glacier. It rises as an immense dome up to the foot of the Rimpfischorn. The ice appears to be benign, its slope gradual, and we continue unroped. As we ascend, stellar snow crystals give way to feathery hoarfrost that sounds like Styrofoam as we walk over it. The dome is not difficult or steep, but it seems like a long plod upward and does not flatten out on top, as expected, but steadily lessens in grade up to the lower rocks of the mountain. There, where the glittering white of snow meets intricate dark gray of rock, we stop awhile to breathe.

We ascend cautiously up the snow-covered wet rock, glad to have not yet put on our crampons. The rock, darker and greener with abundant phenocrysts, is beginning to look more like metamorphosed peridotite, which only appears on the Earth's surface as a result of particularly violent bulldozing and uplift of deep sea floor material. Its widespread presence here indicates that such events indeed took place in the formation of the Alps, and a quick look around at the vertical relief of the terrain also testifies to the thesis of dramatic tectonic forces at work. The rock proves solid for climbing on, and soon we begin to enjoy the physical activity, done with the monotony of walking continuously uphill on snow. We ascend several hundred feet, using our hands on the snow-covered rock where we often would not on a dry summer day. Finally we pull ourselves up to a second, low snow dome. Immediately before us towers upper Rimpfischorn, seven hundred vertical feet of black stone battlements and hanging snow and ice, its jagged uppermost reaches wreathed in dancing, silky clouds. If there was any doubt in our inspired minds, it is cast aside. This is a real mountain.

We crouch in the wind and snow, tinkering with our crampons with cold, wet, ungloved fingers. I crane my neck up towards the summit. It looks like the dark stuff of the rock band below. Deep sea floor basements all the way up here at thirteen thousand feet. My mind flashes back to the limestone walls and peaks of the Bernese Oberland. Limestone. Formed in shallow seas. It occurs to me that the whole of the Alps is an uplifted, twisted, folded, refolded, cooked, crunched, deformed, magma-injected ancient sea floor sandwiched between vagrant Italy and the continent. But the Alps are just one link in a greater chain of mountains that extend all along the underbelly of Eurasia, including the Pyrenees, the Alps, the Taurus, the Caucasus, the Zagros, the Karakoram, and the Himalaya. These are the mountains of the Alpine-Himalayan Belt, and they form the highest and the most dramatically vertical terrain in the world.

If the nineteenth century was the golden age of mountaineering, the twentieth century was the golden age of mountain geography. Over the last hundred years, as the last unexplored continents and ocean basins gave way to the tracks of skis or the keels of ships, the Western world turned its attention ever more to the mountains. Rugged, remote, and inaccessible, the far-flung mountain landscapes of the world were (and still are) among the last landscapes to be mapped, studied, and understood. But initial explorations and mapping efforts were superficial, and the secrets of the origins of mountains beguiled the best minds of the west. Even as the highest peaks knew the ax and the crampon, we knew precious little about them.

How and why did they form? How did glaciers influence their topography and uplift? Why were there fossils embedded in sea floor sediments two miles above sea level? Why did mountains occur where they did?

The theory of plate tectonics was to geology what Darwin's theory of evolution was to biology. It revolutionized mountain geography like nothing else had or probably ever will, and put mountains center stage in our understanding of the Earth's surface processes in general. But the theory didn't develop overnight. It was decades in the making, and required the contributions and expertise of dozens of individuals in different fields. Unlike many of the breakthrough theories of science before it, the development of the theory of plate tectonics was a collaborative effort. Briefly, the theory asserts that heat generated deep within the Earth causes planet-scale convection cells in the sphere of the earth that occurs between the core and the crust. These cells do two things. First, they act as conveyor belts on which relatively buoyant material of the Earth constantly rides towards the outer spheres of the earth (i.e., lighter minerals such as quartz, mica, and feldspar, in addition to the gasses that make up the atmosphere). Second, these convection cells, like upwelling boiling water, cause constant shifting and reorganization of the overriding Earth's crust. As long as heat is generated, thermal convection will keep pumping out the heat, the Earth's innards will keep differentiating according to density, and the crust will keep ripping open, shifting around, and crashing together. Where the upwelling of a convection cell occurs, the crust diverges and is torn open, and rift valleys, seas, and eventually oceans are born. Where the sinking of a convection cell occurs, the crust converges and collides, and mountains are born.

All across the underbelly of Eurasia, forming a wide arc from the Pyrenees to the Himalaya, two convection cells, like conveyor belts, are converging, bringing the crustal plates that include Spain, Italy, Africa, Greece, the Middle East, India, and even Australia into head-on collision with the dominant landmass of Eurasia. The effect is like sliding two carpets toward each other across a smooth floor. When they meet, the leading edge of each carpet buckles up into a fold, and as the collision continues, folds stack up behind the collision zone until the whole of both carpets are rumpled. But it's not quite that simple. Before the continents collide, the ocean basins between them must be squeezed out of the way.

Not all crust is the same. The crust of the continents is relatively buoyant, made up primarily of crystalline basement rocks like granite and gneiss, overlaid with a veneer of sediments, which tend to build up in thickness unless glaciers come and scrape them off. The material that makes up these rocks was once deep within the earth, and over billions of years has differentiated out towards the surface, rising as magma with heat-driven convection cells and crystallizing at or near the surface. Once the stuff of the continents crystallizes, it is there to stay. Thus, continental crust is ever being built and added to, but only very rarely being taken away. Oceanic crust, in contrast, is relatively dense and is made up primarily of basaltic rock, overlaid with a veneer of sediments that are washed into the ocean basins from adjacent land. These sediments tend to form neat, horizontal strata over the sea floors, like those strata now famously exposed in the Grand Canyon and the Colorado Plateau. The material of the oceanic crust also differentiates from deep within the earth, moving as magma along heat-convection cells, but the sea floor basement rocks are made up of darker, denser mineral constituents than their continental counterparts. Because this basaltic oceanic crust is relatively dense, whenever it collides with more buoyant continental crust, it subducts, reenters the mantle, melts, and is recycled. In contrast to long-lived continental crust, oceanic crust is relatively short-lived.

When continental and oceanic crust collide, such as is occurring all along the western margins of the Americas today, the more dense oceanic crust subducts beneath the encroaching continent and reenters the semimolten interior. As it subducts, it often scrapes the less-dense material off the underside of the continent and mixes with it. This new mix of semimolten oceanic and continental crust, like a balloon pushed under water, is too light to stay deep, and differentiates outward, back toward the surface. On its way up, as it intrudes back into the solid crust of the overlying continent, it melts more and more of the surrounding country rock. The more continental material it integrates into its semi-molten body, the bigger the magma chamber gets, and the more it rises. Enough of this will cause the magma to extrude out of the Earth's surface, resulting in extensive volcanic arcs, such as the Cascades and the Andes of today, just inland from subduction zones. Beneath these volcanic arcs, huge reservoirs of magma collect and solidify below the earth's surface, forming vast bodies of granite that add to the bulk of the continents.

Three hundred million years ago, during the Paleozoic geologic era, the southern continental landmasses, which had long been consolidated into the supercontinent Gondwanaland, began breaking up. South America and Antarctica rifted from Africa, India, and Australia, and headed west. India and Australia began their own migrations, Australia moving opposite Antarctica, and India heading north toward Asia at record speed. By sixty million years ago, during the early Cenozoic era, rifts that had formed in the Indian Ocean basin spread under east Africa and began rifting the continent. Meanwhile, chunks of north Africa rifted from the mother continent and began closing in on Europe. The Old World was on the move.

Throughout all of these continental movements, the seas that once separated the continental landmasses have since been devoured, and new seas have opened up. For hundreds of millions of years, Gondwanaland had been separated from the landmasses of the north by the Tethys Sea, which formed a widening arc from east to west between Africa and Eurasia, and opened to the proto-Pacific. As India and pieces of Africa began their northward grind toward Eurasia, the sea floor of the Tethys was devoured.

Unlike today's situation along the western margin of the Americas, during the early stages of the Alpine-Himalayan mountain-building event, the sea floor of the Tethys was being subducted from *two opposing sides*. Not only was the Tethys forced to submit to the overriding encroaching landmasses of Gondwanaland origins, but it also dove beneath overriding Eurasia. As the landmasses of the south drew nearer to Eurasia, mountains of Cascadian, perhaps even Andean proportions built up at the leading edges on both sides of the Thethys. Island arc volcanoes and metamorphic accreted terranes of uncertain origin stacked up along the underside of Eurasia, the north coasts of India, and the vagrant bits of Africa. It was as if two bulldozers were bound for a head-on collision, each piling up increasingly larger mountains as it bulldozed its way along. The bulldozer to the north was much larger and steadier, while those from the south were smaller but moving dangerously fast. Eventually, the landmasses and their associated mountains were bound to crash.

The Middle East met Eurasia, and the Zagros, the Taurus, and the Caucasus were born. The Tethys was pinched off, and the Mediterranean was born as a shallow remnant basin separating Africa from Eurasia. Initially, it extended through the Black, Caspian, and Aral Sea depressions, but soon the Mediterranean was divided, and the inland seas isolated. In the new Mediterranean basin, the islands of Corsica and Sardinia were torn from Spain and France and relocated offshore from an embryonic Italy. More pieces of North Africa broke off and collided with Europe, pushing up into the underbelly of the continent, and the Alps were born.

The place where two originally separate continental landmasses meet is typically indicated by the presence of old, dark, dense oceanic rocks from the underside of the sea floor basement. Their presence indicates the epicenter of collision, where convergent forces were so great they squeezed up the sea floor basement, turned it on end, and thrust it upward many thousands of feet. Rocks such as ophiolite and peridotide, such as are found on the Rimpfischorn, are not uncommon here. On either side of this suture zone, the crust, like the colliding carpets, rumples, folds, and contorts outward from the suture. The folded crust may be bulldozed sea floor sediments (such as in the Jungfrau region), it may be volcanic island arcs (less common in the Alps largely due to extensive glacial erosion), it may be the exposed granitic basements of the arc volcanoes (such as in the Mont Blanc region), it may be accreted terranes (throughout the Alps), or, in extreme cases, it may even be the original crust of the continent. In almost all cases, the crust is radically deformed, pushed back upon itself in immense recumbent folds, or nappes, or in gigantic, broken thrust faults. Imagine two trains colliding. When they hit, the engines meet in a violent embrace, but everything behind stacks up in a jumble and can end up just about anywhere.

The Alps are growing today as much as they ever have been in the past. The oceanic crust of the Mediterranean is presently subducting beneath Europe as Africa grinds its way north. Thousands of years from now, the Strait of Gibraltar will close and the Mediterranean will dry up (as it has several times in the past). Millions of years from now the entire basin will become extinct. Africa and Europe will become one. The Alps will continue to grow, and will eventually be overwhelmed by a hoard of new mountains that will rise all around them. Or perhaps the Alps will rise above, glistening white and spire-like against a cobalt sky.

On the Rimpfischorn, the first section of snow is steep and soft, exceeding fifty degrees. But underneath the new layer is the hard snow of summer. Our crampons bite in and we ascend, plunging our ice ax shafts in at an angle to aid our climbing. Two or three hundred feet of this, and we traverse left and onto a snow-covered rock rib. We work our way up carefully but with a high degree of confidence, our crampons screeching and scraping against the old, dark stone of the Tethys sea floor. We follow mixed rock and snow, which spirals us upward from the west to the northwest face. Here, a second section of steep, soft snow leads us back right and to a higher point on the rock rib.

In normal late spring or summer conditions, the climb would seem relatively easy, but today we find ourselves second-guessing our decisions and wondering if we should slow it down, rope up, and protect the route. But thus far things have gone smoothly, so we save our snow pickets and continue upward. We step around to the left to avoid a steep section of snow-covered rock, and we are faced with another steep, soft-snow section. It is a short section, but below it the run-out is steep and the mouth of a small crevasse is visible. But we are feeling good and moving efficiently, and the swirling clouds that periodically envelop us have us all too aware that our window of opportunity is shrinking. Given the secure snow beneath, we go for it, unprotected, and climb back up to the rock rib.

The snow brings us up to the crux of the rock. It would be no problem on a dry summer day, but alas, summer has left us. What would on other occasions be an easy scramble is now a long series of steep, exposed moves on slanted, snow-covered holds made still more precarious by ever-screeching crampons. We keep moving, and soon, thankfully, the angle lessens, though the exposure does not. Unflinching as of yet, we pull ourselves up to the southern summit, a narrow snow-covered spire of ultramafic rock, perhaps the very suture of Italy and Europe, standing boldly upright amid the swiftly moving silver clouds.

A narrow col, spanned by a snow catwalk no more than 150 feet long and 5 feet wide, dropping precipitously east to a vertical rock face and west to a sixty-five-degree snow slope, is all that separates us from the final scramble to the holy cross-topped northern and uttermost summit of the Rimpfischorn. We stop for only a minute, breathe, and cross slowly, one at a time. The clouds roll back, and the summit is bathed in pure light. We quickly ascend the short section of rock to the top.

Breathless, we look around us, and the world drops off on all sides to gnashing black rock faces and curvaceous glacial headwaters. It is a dramatic summit, a proud mountain on this day, and we are fortunate to be able to share in its space. But our stay is not a long one. Perhaps only minutes. Clouds hang over us, obscuring the south summit, flying to this point but not from it. The anticipation of the downclimb begins to tear at us. The tragedy of the summit strikes us with a vengeance, and we cannot relax here, knowing that the hardest and too often most fated part of the climb is still incomplete.

As we descend, the thought of roping up and exchanging belays is ever on our minds, but the rock offers few reassurances in the form of good anchor placements, and the snow on top is but a dusting that will not accept our pickets. I reflect for a brief moment on Whymper's descent of the Matterhorn. Swiss guides almost always rope up with their clients, but the practice seems a false security without protecting the route, and as Whymper's tragedy made clear, the consequences can be fatal. We continue down unroped, scraping down the difficult rock section, groping for adequate handholds to lend us the security that our feet cannot. Soon we find ourselves at the uppermost steep snow section, and I point my feet sideways and down and trust the hard layer beneath the powder. My heart leaps twice, but I arrive at the bottom safely. Mike, behind me, meets with greater difficulty, and is stopped at the top of the slope. He looks down into the white cloud and I watch the expression on his face as he recalls the crevasse down there. I curse myself. He has the pickets. I have the rope. Being in the lead, I should have remembered the difficulty of the slope. We should have stopped and roped up; we could have protected this descent easily. I offer to come up and give Mike a belay, but he declines. He doesn't move. I step over to the bottom of the slope, about to head up, and he takes his first step down. Then he takes another, and another. I'm ten times more scared watching Mike than I was descending the slope myself. He makes it down. I apologize profusely. Both a little shaken, we stop for a rest.

Below our resting spot, the descent proves easier. We screech over the rock rib with little difficulty, plunge-step down the middle section of snow, and scrape across a final section of rock. We look down the final steep snow apron and see where it meets the snow dome at the foot of the mountain in a gentle, smooth, graceful curve. Giddy, we glissade the final two hundred feet, powder flying all around us.

On the snow dome we stop to remove our crampons, strap our ice axes to our packs, and prepare for the long glacier walk back to camp. Today we climbed into the very heart of the Alps, to the uppermost heights of that place where the crust of Italy and that of Europe met for an orogenic kiss. But today the world is quiet, and the great movements of the earth seem like distant echoes from beyond the clouds of time. Looking around, we realize our route is obscured by a thickening mist. We follow our own footsteps downslope, and the afternoon clouds envelop us.

Rimpfischorn

# VI

The glaciers here are the largest and most extensive I have yet spent time on. Even so, they are but a distant echo of their forebears, the glaciers of the massive Alpine icecap of the Pleistocene epoch, which reached far down into the Swiss Mittelland, and deep into Italy, Austria, France, Germany, and Liechtenstein. Today, only the tributaries of such an expansive system of glaciers remain. But still they are worthy of respect. They cover hundreds of square miles. Many of the larger valley glaciers reach far down into the Alpine and even sub-Alpine pastureland, and many a belled cow munches herbage at their edges. Hanging glaciers send down hourly seracs and avalanches to the waiting valley ice below, yet still they ablate, and the majority of the valley glaciers bear the telltale sloped, sinking, debris-covered snouts of ice that is dying.

It rained last night, and as the air temperatures dropped the rain froze and eventually turned to snow. Morning temperatures linger in the twenties, and I am reminded of the recipe for a glacier—it is quite simple: cold and wet. We have had plenty of both, and this morning I have a difficult time believing that the Alpine glaciers could be doing anything but growing this year. We pull ourselves from our sleeping bags and out into the clear, fresh day. We pack up our camp with cold, bare hands, and stuffing sleeping bags into stuff sacks quickly turns our fingertips to numb stumps. We have to stop frequently to warm up, and every time we do, the feeling of blood returning to my hands makes my fingers throb painfully.

By midmorning we have set out across the slick talus, down the valley toward the Fluealp hut. Soon we leave the dark sea floor sequences behind us and the terrain widens to meadows, yellowing but still green, broken by rock outcrops and small, abandoned lakes. Rising above the yellow-green of the autumnal vegetation, the lonely spire of the unmistakable Matterhorn juts up into an otherwise unbroken clear, blue sky. Its form is remarkably higher than it is wide, its flanks fifty degrees at least, and a slight crook in its otherwise symmetrical form somehow makes it even more perfect, and real. There are no other summits in its vicinity, and the blankness of space on either side of it make its imposing form still more bold, stark, and clean. In ambience, it is the Grand Teton times ten. No, it is unlike any other mountain.

In the foreground, a group of at least twenty chamois graze comfortably in the meadow. There are many young among them, and I wonder which are this year's and which are yearlings. They do not scatter or even move at the sight of us, but go about their business munching on grasses and herbs. I stop and watch them for a while, crouched behind a rock. After some minutes, I can't stand it any more. I slowly emerge from my blind and tiptoe toward them. The chamois are not fooled. They nonchalantly shuffle away, over a rise, and out of my sight. Feeling primal, I drop my pack and decide to chase them. At breakneck speed, I head for the rise, moving quickly but making little noise. The wind is in my favor. As the slope below me comes into view, they are there, and by the time they see me their heads are cocked upward and their ears erect. They turn and flee. Before they totally disappear, a single animal pops its head up over the slope for one last good look at me. I imagine myself an Ice Age hunter. I think I could have had one of them.

We continue down to the small tarns blocked in by the steep lateral moraine of the Findelgletscher. There, on the rocks that fringe the tiny lake, we strip down, and before we can convince ourselves otherwise, we jump in. The water is freezing, in the upper forties at best. We both take a quick dive, emerge, and pull ourselves out onto the rocks, gasping and laughing nervously. We do not wait for the fickle sun to dry us, but rather dress quickly, goosebumps forming on our skin in the cool mountain wind.

We scramble up the loose, unsorted rock of the lateral moraine and onto its top, where a worn, narrow footpath follows its spine downslope. Here, the dwindling form of the Findlegetscher comes into view. It is a debris-covered, elongated, depressed tongue, shrunken in a bed of seemingly oversized lateral moraines. The moraines form long hills of unsorted, unstable rock alongside the glacier, and these hills pile up to abnormally high proportions, further accentuating the sunken form of the ice below. But as we look up the glacier toward its headwaters, its form gradually fills out, until eventually it spills out of its confines and mingles with the headwaters of the Gornergletscher to the south. Above, more snow falls on the glacier each year than melts, and the ice accumulates. Here, the melt exceeds accumulation. If the overall accumulation of the whole glacier exceeds the melt, the glacier has a positive economy, and will grow. If the overall melt exceeds the accumulation, the glacier has a negative economy, and will shrink. Even growing glaciers in temperate mountain environments have ablation zones, but they fill their beds, and their snouts are characterized by a steep, clifflike escarpment of ice. The Findlegletscher, in contrast, is shrinking in an excruciatingly obvious way, dying in its bed like some starving, frothing Pleistocene animal well past the point of defiance, its snout so deep in the dirt it is difficult to tell where the ice ends and the ground begins.

The Findlegletscher, like all true glaciers, is a moving body of ice. Like a river, it follows the path of least resistance downslope, and gravity is its ultimate governor. Like a river, it follows the course of existing watersheds, eroding, exaggerating, and eventually transforming the watershed over time. Unlike a river, however, in the case of a glacier the water is frozen, and because it is frozen it can not flow downslope with the velocity or freedom of form that water can. Instead, it builds up and accumulates, fills valleys, spills over passes, and overwhelms the landscape. But it still flows. The mechanics are just a little bit more complicated.

All glacial movement, like the movement of water, is governed by gravity. The rate of movement may vary greatly according to the slope angle, temperature of both the air and the ice, and the bedrock topography and structure of the underlying landscape. Typically, mountain glaciers such as those of the Alps flow at a rate varying from ten to a thousand feet per year, translating roughly from a fraction of an inch to several feet per day. Movement is typically slow and gradual, not something to be witnessed in an hour's sitting. Occasionally, however, larger valley glaciers will surge, flowing up to a hundred feet per day. Such surges occur most often in response to climatic fluctuations, and they are well documented today among the glaciers of the Alps, as well as those of coastal Alaska and other mountain and maritime regions. Even relatively recent history is punctuated by tales of entire Alpine villages being subsumed by surging glaciers, buildings, houses, and in some cases inhabitants wiped out by the abruptly encroaching ice.

Gravitational stress put on glacial ice will cause it to respond in one of three ways. *Basal slip* occurs when a glacier literally slides downward over the underlying bedrock. Such slippage can be catastrophic and is the main cause of surges. This kind of movement is common among mountain glaciers from tropical to subpolar latitudes, where a thin layer of pressurized water occurs between the ice and the bedrock, acting as a lubricating agent. Basal slip is much less common among colder polar glaciers, which are frozen to their beds. Short and strong stresses, such as those resulting from earthquakes, will fracture ice along planes, resulting in a kind of movement called *shearing*. Shearing occurs when sheets of ice move over other ice along these fractured planes, much like a stack of loose paper placed on an incline. Long, mild stresses, such as the continuous pull of gravity, will cause ice to deform. Ice may stretch elastically to accommodate the additional stress, or it may flow plastically, in a solid state, when the internal stress exceeds the elastic limit of the ice. Such *plastic flowage* occurs by granular and intergranular shifting, and continuous recrystallizing. The ice crystals glide along their hexagonal planes by melting and refreezing in response to differential pressure, slowly moving in the direction of gravity's pull. This transfers the glacial material downslope like a creeping river. Such flowage will increase as ice thickness and slope angle increase. Plastic flowage is the primary mechanism of movement for most large glaciers, including those of the polar regions, though it also occurs in mountain glaciers, such as those of the Alps.

Glacial movement occurs differentially throughout the glacier. In general, like a river, the middle part of the glacier tends to move faster than the sides. Velocity also changes with depth, and as a glacier moves over varied topography, different parts of the glacier move at different rates. To flow over a hummock, for example, the glacier must accelerate and thin, just as a river accelerates and becomes shallower as it moves over a boulder. The deepest ice responds to this by either elastic or plastic deformation, but the upper ice has much less pressure on it and so cannot deform, or metamorphose. Here, the brittle ice cracks, forming deep crevasses in the surface of the glacier. Crevasses may well exceed a hundred feet in depth, though those of smaller cirque and valley glaciers are rarely so deep. For much of the year, their yawning openings are covered with snow, and it is typically only during late summer that they are open wide for the looking.

Several times over the last two million years, massive mountain icecaps accumulated and flowed, slipped, and sheared their way down the flanks of the Alps. The numerous advances and retreats of the Alpine icecap during this time correspond directly with the advances and retreats of the great continental ice sheets of the north, which overwhelmed Scandinavia, the Baltic, the North Sea, and most of Britain and Ireland, as well as a third of the North American continent. The exact number of advances and retreats is not known, as each successively larger advance tends to wipe out the evidence for smaller advances that may have occurred before. Estimates usually hover around eleven major advances, but some scientists advocate for at least twenty. While the number of advances is not certain, the fact that we are in the midst of an interglacial retreat is. The ice will return; it is only a question of when.

Each time the glaciers of the Pleistocene advanced, the ice-free land in Europe was sandwiched between the northerly continental ice sheet and the spreading mountain icecap of the Alps. To comprehend the extent of Pleistocene glaciation in the Alps, take the glaciers of the Icefield Ranges of the Alaska-Yukon border country, the largest nonpolar glaciers in the world, and multiply their extent by two. The resulting sandwich of ice-free land severely restricted the amount of habitat available for plants. Over at least eleven such episodes, the biological diversity of Europe plummeted. Southern refugia was either blocked off by the Alpine icecap, sealed off by the Mediterranean basin (which was evaporating at the time due to lower sea levels), or limited by intense competition with the plants that already had rootholds to the south. Even today, the number of species of plants and animals in Europe pales in comparison to that of North America. According to some sources, there are more species of plants in the Great Smoky Mountains National Park alone than in all of Europe combined.

It would be difficult to overstate the influence that the glaciers of the Pleistocene had on the shaping of the Alps. Each time the glaciers advanced, they scoured, scraped, and quarried the mountain landscape, digging drainages deeper, steepening valley walls, flattening their bottoms, plucking away at cirque headwalls, and undermining the already steep cliffs above. They acted as gigantic rasps, sharpening the peaks and carrying their extensive quarry, those rocks within the ice and those trillions of tons that had fallen on top, down, down, down to the washed-out valleys below. Every time the glaciers retreated, they did so in an incomprehensibly large-scale slurry of silt, sand, gravel, stone, and even boulder-laden meltwater, a flood of epic proportions, further scouring valleys, filling up their floors with freshly gnashed material, washing out sand and dust to the low-lying surrounding regions and even to the sea. With each and every cycle, whatever meager soil had accumulated since the last episode was scraped off, again and again, only to slowly, tenaciously reclaim the newly exposed barren landscape upon the next retreat.

The repeated erosion of the Pleistocene, still at work today, sculpted what was likely a high, plateaulike mountain landscape dissected by deep V-shaped river valleys, into the world of arêtes, horns, spires, cols, U-shaped valleys, and hanging waterfalls that the Alps are famous for today. Broad ridges, the flanks of which opposing glaciers gnawed away at as they eroded headwardly, were worn away to jagged, bare arêtes, and some eventually to cols. Where bedrock was most resistant and did not give way so easily to the wreckage of ice, as in the Mont Blanc region, these arêtes were lined with needlelike spires, or aiguilles. Where glacial erosion was still more extensive, and where three or more cirque glaciers found their heads on opposite sides of a massif, the ice carved horns such as the Weisshorn, the Shreckhorn, the Matterhorn, and countless others.

As the ice flowed down from the highest mountains and into the established valley systems below, it collected and gained volume, filled and often overfilled the valley rims, spilling over divides and forming vast, interconnected icefields. Below these thick masses of ice, deep within the cold, blue lightlessness of the glaciers, the bedrock yielded to the constant downward, outward pressure as water crept into the innermost cracks in the stone, froze, and wedged it apart. Gravity did the rest, and the quarried stone, suspended in slowly moving blue ice, began its long journey to the low country.

All of the material the glaciers quarried and eroded from the mountains was eventually transported, either bulldozed in front of the ice, or conveyed in or on top of the ice to the adjacent lowlands, where it was redeposited in enormous moraines, glacial erratics, and other assorted till. All this removal of material, including the release of overlying pressure from the ice itself, lessened the burden on the deep, underlying crustal root of the Alps. Where trillions of tons of rock and ice had been pushing down on the root for millions of years, now, relatively suddenly, that weight was removed. Like glaciated regions around the world, including the Himalaya, the Caucasus, numerous ranges around the Pacific Cordillera, as well as the Baltic and Canadian Shields, the Alps experienced pronounced isostatic rebound, or uplift, throughout the Pleistocene, as a result of the removal of overlying material, including both rock and glacial ice. That uplift, along with the original tectonic uplift, is still going on today.

The last ten thousand years have been no exception to the rule of erratic climate changes that have characterized the past few million years. Since the Climatic Optimum of six thousand years ago, global temperatures have generally cooled, but not without a host of relatively minor fluctuations that have contributed considerably to the development of Western civilization. These fluctuations bear names reflecting such influence: the Climatic Optimum, the medieval warming period. Perhaps the most notable of these fluctuations occurred between 1300 and 1850, the period known as the Little Ice Age. The general trend throughout this period was, again, cooling, but the five-hundred-year span is also marked by devastating droughts, deluges, crop failures, bumper crop years, and record high temperatures. Not coincidentally, it is also marked by the Age of Exploration, when Europe looked toward wider horizons for resources in an age when their own became unreliable. Meanwhile, the colonization efforts of the previous warming period, such as the Norse Greenland colonies, were overwhelmed and ruined by permafrost, cold summers, and swarming sea ice. In the Alps, entire villages were abandoned before advancing valley glaciers, which unflinchingly swallowed the emptied human structures and continued down valley. Some villagers were not so lucky, and gave their lives to the encroaching ice. For a while, it must have seemed that the whole of the Alpine region would once again be inundated in a vast mountain icecap.

While climatic fluctuations since the retreat of the Pleistocene glaciers have been erratic, they have not been without pattern, and it is seldom doubted by the scientific community that were are in the midst of yet another interglacial retreat, and the great ice sheets and mountain icecaps will return. All of the major astronomical, geological, and meteorological factors that contributed to the development of the Pleistocene Ice Age (and there are many) are still at play; it is only a few minor factors that, at present, are not. It is not a question of whether the Pleistocene glaciers will come back, but when. The Holocene, that epoch of geologic time assigned to the present, is a myth of human making, and the last ten thousand years of relative warmth have done little to subvert the dominant paradigm of the Pleistocene.

The Alpine glaciers of today are at a historical all-time low. Probably not since the Climatic Optimum, and definitely not since the Medieval Warming Period have the glaciers been in such a shrunken state. In the lower reaches of the Alpine glaciers, those moraines that seem too big for the sunken glaciers between are Little Ice Age moraines, and clearly mark the extent of ice from those centuries not so long ago. When Whymper climbed the Matterhorn, these glaciers all filled their beds, and in many places a hundred vertical feet of ablation has occurred in less than two hundred years. A visit to almost any mountain glacier of the midlatitudes, and many glaciers of both equatorial and polar latitudes, will tell the same story: shrunken tongues of ice with telltale Little Ice Age moraines loosely hemming them in, chunks of disjunct dead ice below the terminus, covered in debris, like snow piles in a parking lot in spring holding out just a little longer before the big melt. Recent research reports indicate that as much as a third of the Arctic sea ice of the last millennium has melted since the Little Ice Age. In the last decade, Russian ships have made the North Pole and found open water. Permafrost across the circumpolar north is in decline and the temperature of frozen ground on the rise, with active layers increasing and Arctic villages relocating as the ground of coasts and islands melts for the first time in millennia and oozes seaward. Chunks of ice the size of small nations calve off the Antarctic ice shelves. There is no longer even a shadow of a doubt in the scientific community that the abrupt human-induced reintroduction of greenhouse gasses (most notably carbon dioxide and methane) since the onset of the Industrial Revolution is a leading cause of global warming, and a catalyst for global-scale catastrophic climatic change. Average temperatures in northerly regions are six degrees Fahrenheit higher than in the past, and rising. Repeated record-breaking high temperatures highlight climate statistics worldwide, accompanied by drought, insect infestations, changing ocean currents, and generally anomalous weather patterns across the globe. As the world gets warmer, the fate of ice on Earth is sealed. The glaciers are dying.

The descent down the lateral moraine is like sliding down a scree slope except there are basketball-sized boulders mixed in. We plunge-step down crushed schists and serpentines and trust rounded boulders only loosely fixed in place, but the whole slope moves with us in a gravelly rush. We dodge the big stuff and keep sliding down, and eventually the angle lessens. Here, small streams emerge from the till, and I begin to dig into the debris. It is wet, and soon my arms are soaked up to the elbows. There is ice in there, old ice from better days, covered with a foot and a half of debris that insulates it from the summer sun. But the sun seems to be having its way, and meltwater seeps out of the slopes of the lower moraine in a score of tiny streams.

We step out onto the glacial ice. It is coated by a veneer of debris and sand, which has melted slightly into the surface of the ice, making a texture something like ten-grit sandpaper. It is easy to walk on, and we cross the width of the glacier a few hundred feet up from its terminus, where huge volumes of turbulent water can be heard gushing out of the ice. Here, so close to the snout, the glacier almost looks more like rock than ice. It has the smooth, rounded shape and light gray color of glacially polished granite, and from afar it might even be mistaken for such. But as we move down closer to the snout, the angle of the ice steepens, and there are large cracks. From inside the cracks, we can hear water, just a few feet below the surface of the ice, moving between it and the bedrock. The cracks extend downward, following the depressed contour of the terminus, and at the edge, frothing, milky, gray water comes leaping out of the confines of the ice in seething little torrents. These quickly merge, braiding into a booming, hissing meltwater stream.

I cautiously step toward the cracks and find an ice ledge that leads down into the glacier. I carefully step onto it, and there is just enough debris to provide good footing. It goes down maybe five feet, and ends next to a blue hole. Peering into the hole, I watch meltwaters ten feet beneath me twist and convulse among ice and boulders, looking for a way out. The contrast between the translucent, blue, still ice and the opaque, pale gray, churning water is made evident by this scene. The fact that the one turns into the other requires some reckoning. I stand there for a long while, amid the rush of water, listening to the blue ice booming. It makes the rest of the world seem silent.

Cold, I creep back out of the cracks, repeatedly wiping the fine mud and sand of the ice's surface from my hands. Soon the legs of my pants are covered. Mike is standing at the terminus, watching the water spill out, and he is quiet and still. I walk toward him, over a dozen clear, cold runnels—small meltwater streams streaking their way across the surface of the glacier and leaping off the clifflike face of the ice and into the larger, milky braids below. I follow them down to gravel bars that are dissected by streams. Here, detached from the terminus, truck-sized towers of dead ice stand in place, coated with protective debris, dripping in the afternoon light. Mike climbs partway up one, and on the slope he runs in place, the comical expression on his face like that of a circus performer, demonstrating that indeed, under the disguise of black debris, the entire mass is of ice. Its fate is sealed by its isolation in a warming world. It will sink into the soggy ground of the outwash, melt into the gravel, and become a sapphire kettle lake gleaming in the summer sun.

Gornergletscher

# VII

Rain since midnight and no sign of its waning. We sleep in, and in, and in, and make no move to budge. The mountains are hidden by clouds and our world has shrunken to the buttelike massif of the Rif-flehorn and a square mile of visible alpine pasture land. Snow is falling on the high mountains and the end of summer is at hand. We have come here to say our goodbyes to the Alps, to give our thanks through one last night on the ground, and one last day in the open.

Away from the train tracks, tucked behind a small, low hill and the cliffs of the Rifflehorn, is a shallow, placid lake. It is surrounded by flaky serpentine gravel chunks that over the years have arranged themselves flat-wise, like a broken mosaic or old, cracked pavement. The flat stones prove to be a good bivouac, and the mirrorlike pool eases our minds. Even in the wind, when the rain stops, the pool becomes smooth and glassy in a matter of minutes and its reflective qualities are impressive. But it mostly reflects shades of gray.

After several cups of tea and much lying around, we finally get up and out. We spend the better part of the day scrambling on the slick rocks of the Rifflehorn, finding corner systems and cracks and chimneys on its north face and climbing them to the summit. There, we sit amid the clouds and wonder where the mountains have gone. It is somber up on top. We don't talk very much.

The hours slip by. The fact that this is our last day in the mountains doesn't seem to grip us. We each walk around the massif of the Rifflehorn alone, through the rain, out of the rain, and back into it again. Occasionally the clouds thin and lift, and below the Rifflehorn to the south, the immense ice of the Gornergletscher is unveiled. This, the largest valley glacier in the Pennine Alps, has blue rivers in ice canyons a hundred feet deep flowing on top of it. Numerous lobes of ice descend from the cloud bottoms, steeply pouring their broken blue, gray, and white masses into the valley, where they merge with the telltale black stripes of medial moraines onto the master glacier. Above, invisible, cloaked in an impenetrable mist, is the sturdy mass of Monte Rosa, the second-highest peak in western Europe, the highest border between Switzerland and Italy, the source of the ice. But today its countenance remains obscured, and if not for the enormity of the Gornergletscher below, I would doubt it was even there.

Back at camp, an imperceptible early evening sets in, and though we would not know it from any change in the dim light, our stomachs tell us it is supper time. The rain stops as we cook, and we are glad to be able sit out on the shores of the reflecting pool and look at the world in clean light for a while. Nearby, a lone ibex clinks among the rocks and nibbles herbs. It is an old beast, its horns gigantic, obviously a heavy burden to bear. Its red eyes glint as the clouds thin, and for the first time ever I am close enough to see its pupils. They are slits, like a cat's. It reminds me of some exaggerated, mythical version of the bighorn sheep of the American West. But this animal, docile, does not run, but stands there peacefully, looking at us from time to time, and goes about its business.

Pale evening light increases its glow as the clouds thin, and patches of blue appear and disappear above as the air cools and becomes brisk. It occurs to me that when the clouds are in, it feels like the sun is never going to shine again, and when they finally break, no matter how many times I've witnessed it before, I am always astounded, utterly amazed. Mike cracks open the small bottle of whiskey we have been saving for just such an occasion. Soon we are laughing and telling stories, our minds ignited by the combination of malt and clearing skies.

The clouds roll back, fleeing south behind the mountains, revealing at last the glacier-encrusted bulk of the Monte Rosa massif. In the evening light, its form seems indomitable, as if the entire world lay at its feet. And as we are transfixed, the image is doubled, and the reflecting pool, as still and as delicate as glass, frames the immensity of the mountain perfectly and adds to its reflection of the sky the blue-gray outlines of serpentine stones. We become silent. Neither of us expected this. As we sit and watch, the waning sunlight repaints the snowy mountain over and over again, first with a silver brilliance on an azure sky, then with amber and peach as the sky deepens to purple. The image burns itself into my psyche: the image of the magnificent Alps, clad in snow, burnished with the changing light of the evening sun. Scenes flash before my eyes. There are icemen with grass capes, Celts and Romans, Germanic pastoralists, Romantics, and tragic climbers. There are ancient sea floors and vagrant continents. There are glacial icecaps a mile thick and ice-mantled pinnacles that point like fingers to the heavens. My mind wanders across the continents but I can think of few stories as rich, few scenes as beautiful, as the one that is right before me. I think to myself of a thousand mountains reflected in a thousand lakes, but somehow none are as perfect as this one.

Monte Rosa Massif

# *TEMPLES IN THE SKY*

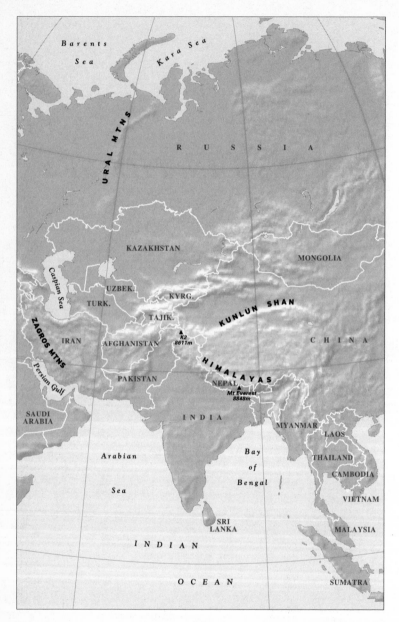

Asia

The Nepal Himalaya

# The Nepal Himalaya

## I

**NEPAL IS HIMALAYAN ASIA** in a nutshell. It is to Asia what Switzerland is to Europe. Both are small, landlocked countries. Both have physical landscapes dominated by high, glaciated mountains. Both are sources of water and sediment that splay out and feed the fields of the surrounding lowlands. Both are cultural headwaters within their respective continents. Both have long histories of independence from their lowland neighbors. Both are microcosms of their respective continents: compact, miniaturized versions of larger landmasses in which diverse cultures and landscapes alike are corrugated and crammed together. Both countries are innovative and outward looking by economic necessity, yet are home to some of the most conservative, traditional cultures left in the world. The fact that these two countries share all of these characteristics is no mere coincidence. In both cases, the single most important factor in the geography of the region is the presence of mountains.

But Nepal, like its European analog, is not entirely mountainous. In fact, standing in the midst of the steaming jungle of southernmost Nepal, just a few hundred feet above sea level, there is no hint of mountains nearby, save the starry-eyed tales of strangely dressed traders from the north. The feeling is of India, the air thick and moist, chattering monkeys in the trees, wandering saddhus in the village lanes. But the mountains build abruptly to the north. Nepal, like a gigantic staircase one hundred miles long and five hundred miles wide, ascends from the subtropical plains of the Ganges, up to the undulating hill country of the Terai, to the steep slopes and deep folds of the foothills, and finally to the highest crystalline peaks and glacier-filled troughs of the Himalaya. Here, the watershed divide of the Himalayan crest forms the border between Nepal and Tibet, and the mountains are crumpled into such astounding vertical relief that the sky seems to back away from the encroaching gray and white loom of the serrated peaks and ridges. And to the north, shadowed from the influence of the monsoon by the high mountains, free of the scouring action of both glacier and river, the arid, treeless expanse of the Tibetan Plateau sweeps with monotypic vastness into the center of the largest continent on earth.

Nepal, though it holds the heart of the range, by no means contains the entire Himalaya. Describing a great arc from Kyrgyzstan to Assam, also taking in parts of Tajikistan, Pakistan, Afghanistan, India, China, Nepal, and Bhutan, the Greater Himalaya span two thousand miles from west to east, separating southern Asia from the bulk of the continent physically, biologically, and culturally. Physically, the fertile lowlands of the Indian subcontinent are bound to the north, where beyond the mountains the landscape is that of a vast, temperate, interior desert. Biologically, the subtropical and tropical species of the subcontinent end at the mountain front, and to the north the vegetation is sparse and prostrate, bending low to the will of aridity, elevation, and exposure. Culturally, Caucasian and Indo-Aryan peoples of the rich heritages of Islam and Hinduism dominate the Himalayan region to the west and south. But in the high mountains, and on the broad plateaus of the interior, these cultures give way to mountain people of Mongoloid origin, Sino-Tibetan speakers, decidedly Buddhist.

The Himalaya are undisputably the highest mountains on earth. Other high mountains, including the Andes, the Saint Elias Range, and the Alaska Range, fall short of the Himalaya by thousands of feet. Including the associated Karakoram and Hindu Kush, the Himalaya contain over thirty peaks in excess of 25,000 feet, and every single

8,000-meter (26,248 feet) peak in the world. By far, the highest concentration of these giants occurs in Nepal, the heart of the Himalaya, including Everest, Kangchenjunga, Lhotse, Makalu, Cho-Oyo, Dhauligari, Mansalu, and Annapurna. Among the high mountain regions of Nepal, the Khumbu region, containing Everest, Lhotse, and Cho-Oyo, reigns supreme. Here, the Himalaya achieves its apex, reaching nearly six vertical miles towards the sky.

I awake to the sounds of cows lowing, bells clanging, and the all-pervasive roar of a rushing river in a deep valley not far away. Mist-tails hang like ghosts of torn prayer flags on the high ridges that loom straight above me, and as I crane my neck upward I can see the warm light of the morning sun dappling the dusty boughs of blue pines. The sky seems small next to the immensity of the definitively vertical landscape, and the pale blue is confined to a narrow space between the mountains. A woman sings in nasally minor tones, a thin wail over the white noise of the river. The air is cool. It carries the comforting smell of woodsmoke. I roll onto my back, eyes wide open, my breath condensing into miniature clouds around my face.

I am surrounded by fields of spring-green barley and mustards. The fields form neat, terraced polygons perched on steep, rocky slopes. These polygons of soil and crops are contained by rock walls that fall to cliffs, which fall to the booming blue-green water of the Dudh Kosi hundreds of feet below. The fields are bound on one side by the drop of the river gorge, and on the other by slopes, or the stone buildings of scattered villages. From the air, the sprouting fields of the lower Khumbu in April form a narrow, unlikely bright green stripe along the deep cleft of the Dudh Kosi, and it looks as if someone took a thin shaving of Ireland and pasted it onto the mountain landscape of the Himalaya. The fact that people farm here beguiles me. Looking at the angle of the slopes, it becomes evident that whatever soil is here had to be caught, carried, composted, or otherwise coaxed into place and carefully manipulated into its present horizontal position, then contained with equal care. Without such terraces and retaining walls, the sluicing rains of a single monsoon season could wash every speck of brown loam to the river below, and down to the plains of the Ganges.

The village here is called Monjo, and we arrived on foot, as everyone does. In our case the walk was from the small airstrip at Lukla, seven miles to the south, but others come through Monjo from distances ten and a hundred times greater, en route to and from the high Himalaya. There are lowland folk carrying bamboo baskets full of chickens in smaller baskets, burlap sacks of rice, spices, and dried chilies. There are highland folk carrying baskets full of potatoes, leaves, and dung-patties. There are livestock trains, beasts laden with mysterious blue and yellow plastic sacks. There are trekkers dressed in awkward-looking western uniforms complete with synthetic zip-off-leg pants and specialized aluminum walking sticks. There are porters carrying their belongings, brightly colored, oversized duffels stacked two high, perched on basket rims, towering over the small, brown heads of their bearers. There are small children in blue clothes, singing and holding hands, on their way to the village school.

The stone lanes of Monjo are lined with rectangular stone buildings, some of which have mortar and some of which do not. Most of them have blue and white trim and face the main footpath, with windows looking out on the regular traffic of passersby. Abundant firewood is neatly stacked against the walls of these buildings, and the smell of wood fires wafts throughout the trees, lodges, and houses clustered between the slope and the river. The buildings back up to the fields, and the feeling is of closeness. There is no room for sprawl here.

I walk over to a large and newer-looking stone building and go inside. The main room is centered around a large woodstove, and the black iron of its surface radiates a comforting heat. Around the perimeter of the room are box-style benches covered with decorative wool mats and pillows, and small, low, long tables. Windows surround the place. Adjacent to the room is an open doorway, and from behind a blanket flap I can hear the familiar sounds and smell the pleasant odors of someone cooking.

The room is empty except for one person. In the corner, clothed in a long blue skirt, a black down jacket, and a scarf, sitting over a steaming cup of milk tea, is River Gates, my traveling companion. We have gallivanted around the Sierra Nevada in each other's company for years, climbing dozens of peaks, teaching natural history and alpine ecology courses, investigating birds and flowers, and discussing the applications of Buddhist thought to our experience of the world. Years ago, we first talked about combining our interests in mountains, mountain cultures, and Eastern thought by going to the Himalaya— the home of rugged mountain people and the birthplace of Buddhism. Three weeks ago I called River and asked her if she was ready to go. Though the call was out of the blue, her response was immediate. The next day we had plane tickets.

River is a biologist. Over the last few years she has studied song-bird populations in western North America. During this time, she has become the quintessential "bird nerd." She regularly gets up at four o'clock in the morning. She sneaks around on her hands and knees, in the bushes, searching for nests. She handles thumb-sized birds on a daily basis. She listens to recordings of birdsongs on her car stereo. She carries a pair of armored binoculars (worth twice as much as my car) around her neck wherever she goes. Her commitment to the avian world sheds a humbling light on my relatively amateur fascination with birds. She keeps me on my toes. We watch birds along the footpaths, and for once I can keep abreast of what's going on. Every bird we see is a new sight. Every song we hear is a new sound. Yesterday, brown dippers by the river, blue whistling thrushes, rufus-bellied niltavas, and a conspicuous yellow-billed blue magpie in the trees. It is as if we have just landed on another planet, or, as is the case, another continent halfway around the world.

I sit down to a cup of milk tea. I am delighted and at first suspect (hopefully) that the tea has plenty of yak milk in it. But after one sip I identify the familiar taste of powdered milk. And there is enough sugar in one sip to sweeten the whole cup. In fact, there is so much milk and sugar in the tea that I can barely identify the flavor of the tea leaves. But I enjoy every sip and after the cup is gone I pour another.

We clear the table in front of us and I carefully unfold our map. Is an old Austrian sheet produced by Erwin Schneider, and being of Germanic make, its level of detail is meticulous. The first time I looked at the map I was astonished. It was the opposite of what I have come to expect of topographic maps of mountain landscapes. Usually the maps are mostly green, indicating the montane forest that cover the flanks of the mountains, with strips of gray and white indicating the rock and ice of the alpine country. But this sheet is mostly gray and white. The few narrow strips of green and yellow indicate the scarcity of forested and arable land. The title of the map explains everything: Khumbu Himal. This is the home of the highest mountain in the world, and as such it is one of the highest mountain landscapes on Earth. Interestingly, the Khumbu region is named for its dominant valley and not its dominant mountain, Everest. This is not surprising. The valley was important to people for hundreds of years before the mountain came into prominence. The valley has always given life, while the mountain has always taken it away.

The Khumbu valley is shaped like a great hand reaching deep into the high Himalaya, its uttermost fingertips tickling the underside of Tibet. Its arm is the deeply incised V-shaped valley of the Dudh Kosi, whose blue-green waters eventually mix with the brown of the Ganges. Moving upvalley, the palm of the hand, at 11,000 to 12,500 feet, holds the three major villages of the Sherpa world: Khumjung, Khunde, and Namche. Namche, through which all traffic must pass going north or south along the Dudh Kosi, is the natural trade center, or commercial hub, of the Khumbu region. From this central area tributary, valleys splay out like fingers toward the northwest, north, and northeast, toward the top of the watershed and the Nepal-Tibet border. These are, from west to east, the Bhote Kosi, the Dudh Kosi, the Khumbu, and the Imja Khola, each separated by prominent north-south trending spur ridges coming off the main Himalayan divide. Where the valleys converge around Namche they are forested, and where there is soil it is arable. About halfway up these valleys, however, the debris-covered snouts of valley glaciers encase the ground in ice, and the growing season is so shortened that efforts to grow crops are abandoned. Further upvalleys, where their profiles sweep upward toward their heads, thick, white, active glaciers pour down from the Himalayan rock, and the mountains are mantled in snow and ice. The heads of all but one of these valleys end in precipitous headwalls covered with heavily crevassed hanging glaciers and are impassible. The upper Bhote Kosi, however, to the west, forms a breach in the wall of the Himalaya, and the ice at its head forms a great, wide, white pathway into the vastness of Tibet.

Ever since I first laid eyes on a map of the Khumbu, the valley of the Bhote Kosi drew my attention with an uncanny magnetism. Something about the Nangpa La, the pass into Tibet, was transfixing. In my imagination, the Nangpa La was an ancient pass, and I imagined centuries-old cultural exchange across its windswept, bitterly cold glacial flats. As I researched the area, I quickly found out that it was closed to foreigners, and armed military personnel guarded the lower reaches of the valley to prevent people like me from wandering the high road to China. On closer investigation, however, I found out that the watch over the upper Bhote Kosi had been relaxed, and that we might travel safely at least as far as the remote seasonal village of Lugare, where a route ascended a high north-south trending spur ridge to an ancient pass called the Renjo La, then spilled down into the upper Dudh Kosi Valley to the east. I knew that the presence of the pass at the head of the valley meant there was a story to this valley, and that its only recent opening to foreigners meant we might see the Khumbu as it once was and in only a few places still is. The valley of the upper Bhote Kosi was like a fly in amber. Part one of a plan was born.

It struck me as I studied the Khumbu that a high route to the east from the Bhote Kosi would bring us farther and farther away from the last vestiges of traditional Sherpa culture and closer and closer to westernized Sherpa culture associated with the last six decades of modern mountaineering and tourism. At one end of such a traverse was the Bhote Kosi Valley, and at the other end the Khumbu and the Imja Khola, both highways used by mountaineering expeditions and trekkers, both highly romanticized and thus commercialized, where one might see at least as many westerners as people of Nepal, where the focus of most travelers is the destination and *not* the journey it takes to get there. And between these two extremes of the Bhote Kosi and the Khumbu-Imja Khola valleys was the upper Dudh Kosi, known to westerners for its commercial hub of Gokyo. Frequented by westerners but attracting only a fraction of the numbers as the Khumbu, I speculated that the upper Dudh Kosi would be a snapshot of what the Khumbu was like decades ago, and what the Bhote Kosi may soon become. A high traverse from west to east, beginning in the Bhote Kosi and ending in the Khumbu-Imja Khola, would be like a walk through time. Along the way we would take in everything from five-hundred-year-old pastoral traditions to corporate-sponsored American mountaineers hauling Honda generators and laptop computers on the backs of Yaks and Asians to the base of Mount Everest. We would spend weeks above treeline, ascend to eighteen thousand feet at least three times, cross glaciers, negotiate snow-covered talus slopes, visit traditional Tibetan Buddhist monasteries, and drink strong chang with laughing locals. At first the idea seemed revolting. Why walk into the bright colors and confused questions of modern culture when we could walk away from it? Why not avoid the Khumbu Valley proper like the plague? But the idea grew on us, and in the end we decided to follow the whims of time rather than to try to make a path against it.

Sherpa dwelling

# II

Prayer stones are everywhere and they are as common along the footpaths as people and trees. Here, they are called *mani* stones. Every stone bears the inscription *Om mani padme om.* As I walk, I repeat the mantra in my head over and over again. Then I do the same in English: *Hail to the Jewel of the Lotus!* Some of the stones are rounded boulders the size of cars. Others are platelets small enough to fit in a hand. Some have symbols three feet high, freshly painted white and black against the gray of stone. Others are carefully carved, the symbols raised, the background chipped away. I figure there must be millions of hours of human attention in all these stones. In places there are tablets heaped into piles ten feet high and dozens of feet long. There are trails worn on either side of these piles such that one could circumnavigate the stones for an eternity. But the intention is to provide passage to the left of the stones from any direction, so the passerby always goes by in a clockwise fashion, an act of piety and respect.

There are prayer wheels too, of all sizes and colors. Invariably, they are cylindrical and spin on a central axis, and are painted or embossed with the same prayer as the stones. Small wheels are usually in rows: gold, red, blue and white, ochre, while big wheels, some bigger than oil drums, are almost always solitary. I spin them all as I go by, and a thousand prayers go out into the universe. I wonder if, after years among them, the Sherpa even see them.

Prayer flags flap from every bridge, post, and chorten. Their uniform appearance, square, blue, red, yellow, and green, frayed at the edges, printed with *om mani padme om*, makes me wonder if they are made here, or arrive in bundles from some dusty shop in Kathmandu. I wonder what they were like traditionally, a few hundred years ago, before industrial times, before synthetic dyes, before mass production. Here, everything is sacred, so nothing is. I wonder if Americans ten thousand miles away, with these same prayer flags hanging from their porches, think about that.

We walk along the dusty footpath that follows the Dudh Kosi, crossing the river twice on steel bridges covered with flags as we travel upvalley. We pass porter after porter, usually followed by a few westerners, and innumerable village people going about the business of the day, each person invariably carrying a wedge-shaped bamboo pack basket with a tumpline across their forehead. The porters often carry a stout wooden walking stick called a *thoklang*. It is the same length and shape as an ice ax, but thicker and used for different purposes. While resting, the porters either sit on the wooden handle of the thoklang or use them to prop up their pack baskets and take the load off their necks for a while. I admire the smoothness of some of the thoklangs; they show evidence of years of use, polished by hand oils over thousands of miles of walking.

The Sherpa women are some of the most stunningly beautiful women I have ever seen. They are usually small-boned with high, prominent cheeks and thick, straight, black hair, and many seem to have a look of determination in their eyes. While the men are apt to wear western clothing, the women are less so; they most frequently wear traditional double-breasted robes and intricately woven woolen aprons, beautiful, durable, and practical. Both their physical appearance and their dress is entirely different from that of the Hindu culture to the south, and the only overlap are the Hindu-influenced brightly printed scarves the Sherpa woman often wear wrapped around their heads. Most of the women I have seen have been indoors, cooking and carrying on business transactions with locals, lowland Nepalis, and westerners alike. But women are also working on the slopes, out in groups collecting leaves and dung in pack baskets, and transporting these fertilizers around the valley. Occasionally a Sherpa woman is seen straining beneath a heavy porter's load, but most of the porters I have seen are lowlanders, and most of them are men or boys.

The Sherpa men are rarely seen indoors in the daytime. During both the spring and fall trekking seasons, a large percentage of the men are employed as guides, sirdars (cooks and camp managers), or porters, and thus spend their days on the footpaths and their nights in camps and lodges far from home. Although they often dress in western garb, they are readily distinguishable from their lowland neighbors by their distinctly Tibetan features. They are mostly lean and sinewy, often a full head shorter than I am, and sometimes even smaller. But their legs, lungs, and hearts are stronger than mine.

Livestock share the trail and are often given more respect and right-of-way than people. They are mostly yak-cattle hybrids, and look like compact, slightly hairy cattle. While the yaks thrive and excel as the livestock of choice above the treeline, down in the middle slopes, hybridization with the longer-lived and higher-volume milk-producing cows of the lowlands is preferred. These hybrid cattle carry goods from village to village, most often in trains of four or more animals closely tended from behind. They also graze freely and independently among the trees around the footpaths.

After crossing the river a second time, we begin to ascend up the steep slopes of the east bank. Here, the Dudh Kosi makes a deep incision into the mountain front, and its downcutting power is revealed in the steep cuts the river has made into the cobbles of its bed and the precipitous limestone walls that rise abruptly from the frothing waters. The faster the mountains rise, the faster and deeper the river cuts, and both the rate of uplift and the rate of downcutting in the Himalaya are unrivaled in the world. The slopes are as steep and rugged as any I have seen, and the overall consistency of steepness is unmatched. Even from here in the middle slopes, the mountains seem to be rising so vigorously that the combined forces of water and gravity cannot keep up. Often the result, rather than small-scale downslope movement of land over a long period of time, is large-scale movement over a short period of time. Catastrophic landslides, questionably attributed to deforestation and agricultural practices of the native peoples, are the norm here. And looking down into the deepening gorge of the Dudh Kosi, I can practically watch the stream downcutting like a knife through butter, into the terrain as the mountains rise.

The forest around us consists of pine and oak. It feels familiar, like so many other pine-oak forests of the Northern Hemisphere. The Dudh Kosi crashes and roars in the background, and for a moment I am transported to the Middle Fork of the Kaweah River in the Sierra

Nevada, halfway around the world, where pines and oaks much like
these grow on the steep slopes of a similar deeply carved river gorge.
But here are blue pines, not ponderosas, and Himalayan oaks instead
of black oaks. And I am at nine thousand feet above sea level, not four
thousand. And in between are thick-leaved, brilliantly flowered red
rhododendrons in the peak of their bloom. Silver firs make subtle
appearances in cool, moist drainages, hinting at the forest to come.

Like all mountain ranges, in the Himalaya, altitude simulates latitude, and ascending the mountains is much like going north in terms
of the vegetation communities one encounters. Here, at twenty-seven
degrees north latitude, the vegetation at sea level is subtropical jungle,
complete with palm trees and bananas. At three or four thousand feet,
the tropical feel starts to give way to familiar temperate hardwoods like
chestnuts and oaks, but the climate is far from what a European or
North American would associate with high mountains. Not until
around eight thousand feet does the forest attain a truly montane feel,
comparable to what might occur at four thousand feet in the Sierra
Nevada, or at sea level in New England, Europe, or northeastern
China. At ten thousand feet the pines and oaks give way to a mixed
coniferous forest of blue pine, silver fir, and juniper, mixed with birch
and scattered rhododendron, and the feel is that of seven thousand feet
in the Sierra Nevada, or sea level in northern Maine, Scotland, Norway,
or Siberia. Between thirteen and fourteen thousand feet, the last strag-
gling junipers bend down to the low, prostrate shrubs of the alpine treeline. For comparison, treeline in Yosemite, at thirty-eight degrees
north, is at eleven thousand feet. In Maine, at forty-five degrees, it is at
four thousand feet. In the Alps, at forty-seven degrees, it is also around
four thousand feet. In Scotland, at fifty-seven degrees north, treeline is
around two thousand feet, as it is in southern Norway, at sixty degrees
north. For contrast, in Ecuador, smack on the equator, treeline is just
above fifteen thousand feet. The upshot: the closer to the equator you
go, the higher treeline is; the further north or south you go, the closer
treeline gets to sea level, until eventually the alpine treeline catches the
arctic treeline, and alpine and arctic tundra become one.

We cross a high bridge over the Imja Drangka and begin the steep ascent up towards Namche. I set out ahead and walk alone, through pure stands of blue pine, rising above the comforting roar of the river to strangely quiet, dusty trails where the ground is parched and the springs are few. I walk past yak trains and groups of Sherpa women collecting dried dung patties. A light wind blows through the loose boughs of pines and makes a hollow, distant sound. But somehow it is comforting, like the sound of the wind in the pines of my backyard, and it dries my sweat and cools me. The switchbacks seem endless, and I know from this morning's look at the map that this will last awhile, so I just keep breathing rhythmically and making steps. Even though I can feel the elevation working against me, it feels good to be working hard, walking off too many days of travel and Kathmandu.

After some thousands of steps, the pine wood opens up and the slope lessens, and I can tell from the foot traffic that Namche is near. I walk past some low rock walls and the village comes into view. Rectangular buildings are built upon terraces that form concentric semi-circles upslope. I stand at the bottom, as if in the center of an amphitheater. A stream runs through the village, under and over makeshift rock walkways, and here, at the bottom of town, a few dozen women and children of all ages wash their clothes and set them out to dry on the rocks. Just past them is a large stupa topped by a white tower decorated with golden eyes. Around its base are rows of prayer wheels, and an ancient woman in traditional dress cleans them diligently with a small, bent knife. I loosen my pack from my shoulders and sit down to watch her. A small boy, maybe eight years old, also sits nearby, and soon we begin to talk in a broken sort of way. His English is impressive and I am humbled. Next to him is a large pack basket, heavy for a lad of his size, and he is having difficulty getting the tumpline onto his head. I grab the pack and hoist it up for him. He slides the tumpline neatly into place, smiles, and shuffles off down the dusty path.

Namche is a critical spot because it is at the natural junction of every tributary valley to the Dudh Kosi. As such, any and all traffic passes through Namche on its way north or south, en route to and from both Tibet in the north and the Ganges lowlands in the south. It is a natural confluence of trade routes, a prime spot for smart commerce, charging tariffs, and double-bargaining. As such, ever since the Sherpa migrated into the Khumbu region from Tibet, Namche has been the cosmopolitan center of Sherpa culture, the place where one can find goods from the corners of Asia, hear tales from lands on the other side of the mountains, and hear half a dozen languages in a day.

I hoist my pack onto my shoulders and follow the obvious path into the heart of the village. The path is worn and dusty and people are everywhere, and soon I find myself walking narrow lanes between open-fronted shops full of a surprising array of Asian trinkets. There are tables covered with prayer beads of many kinds: sandalwood, yak horn, seed, Tibetan, and Hindu. There are bowls of all shapes, sizes, and degrees of décor. There are rings, necklaces, earrings, bracelets, pillow covers, scarves, hats, socks, and sweaters, some made here in Namche, some carried up from the lowlands or snuck in from China. There are shops full of ice axes and crampons, rucksacks, down jackets, and bright red and black parkas bearing the logos for North Face and Lowe Alpine. There are cattle trains in the lanes, along with shop owners, locals, lowlanders, Tibetans, and Westerners, all speaking a variety of mutually unintelligible languages. I am transfixed by all of these scenes and objects but have to be careful not to look at anything too conspicuously lest some expert, opportunistic Sherpa shopowner select me for personal attention. Had the path up to Namche not been well traveled I might be dizzy with overstimulation, but already I find myself getting used to what I would consider crowds back home in the mountains of North America.

In search of rest and a hot drink, I find my way to the porch of a small teahouse and bakery on the right side of the lane, where I sit and order an entire thermos full of milk tea. As I sit and sip, I notice that the clear, warm air of the morning has given way to cloud cover, and here at nearly twelve thousand feet, a close, marbled sky obscures the surrounding landscape. I think back on yesterday, and the pattern was much the same. When we landed in Lukla, in the late morning, the sun was shining and only the highest peaks and ridges were swathed in clouds. By the time we neared Monjo, several hours later, the cloud level had descended until we were in the clouds, and light rain fell in short spurts. Although I cannot see for sure, I speculate that the high mountains are closer at hand than the clouds let on. Perhaps in the morning I will see them unveiled. Still in shorts from the hot, dusty climb up to Namche, I soon get cold, and fumble through my pack in search of my sweater and an extra shirt to wrap around my legs. The tea tastes good to me, and because the air is cold I find myself drinking cup after cup.

Over the clanging of cowbells I hear River yelling up to me from the lane below. She waves and patiently waits for the cattle to pass, then steps up to the porch. She sits down and we order another thermos of tea. By now I am exceedingly excited about everything, even anxious to leave Namche today and walk the three or four miles to the quiet village of Thami. But River reports a headache. We have come too high too fast, and it is time to rest and acclimatize.

Mani Stone

# III

I awake to the sounds of roosters and Tibetan horns. We are camped above Namche, in the yard of the Hotel Sherwi Khangba. The Hotel usually caters to large expeditions, and as we are the only two people at the lodge, our hosts not only ignore us but seem outright put off that we are even there. We make ourselves small. If not for the noise of Namche we would have left this place last night in search of a warmer welcome. But the morning is brilliant, and in my half sleep through the cold night I anticipated the coming of the light for hours. I crawl out of the frosted tent and into the yard, and make my way to a lookout over town.

Behind me, to the east, Ama Dablam looms far above all else, powerful in its bold, spirelike form, dramatically backlit, with straight, white rays of sunlight seeming to radiate in all directions from its bulk. I've only seen Ama Dablam in pictures, and its shape is so distinct that I recognize it at first glance. It is sacred to both the Sherpa and the many Tibetan monks at the nearby Tengboche monastery, and in the lore of this place it is heaped with far more praise than Everest, Lhotse, or Cho Oyo. Although it is not particularly high by Himalayan standards, it stands alone, by far the highest massif south of the Imja Valley.

I pass through the gate of Sherwi Khangba and onto the dusty footpath. Not far down, the terraced fields above Namche begin, dropping down in concentric rings to the blue and green roofs of a hundred lodges and houses. The village still lies in shade, but above it to the west, looming with crystalline clarity in the morning light, perfectly front lit, the hulking mass of Kongde Ri rears up as an immense trapezoidal ridge of rock and snow, looming ten thousand feet above the village. At first, I stare in disbelief, then unfold my map and do some quick figuring. From the Dudh Kosi gorge below, it is over eleven thousand feet up to the summit. That elevation gain happens in less than three miles. The escarpment is steeper than the east slope of the Sierra Nevada. It is steeper than the Alps.

I stand and watch as the sun lights up the face of Kongde Ri, works its way down onto the forested east-facing slopes, then pours over the cold metal roofs of the village of Namche. But the village was up before the sun, and the sounds of business have been in the air for hours now. A hundred hammers can be heard quarrying and shaping stone for the construction of buildings, and the distinctive clinking of metal on rock carries far and wide from the village center. Brash horns emanate from the gompa at the northwest end of town, where a dozen red-robed monks greet the day. Red and yellow-billed choughs caw noisily as they hover over buildings in search of food. A flock of snow pigeons flutters by overhead. A lone Himalayan monal, iridescent blue, yellow, red, and green, bumbles like a glorified chicken in a nearby potato patch. A common raven croaks from a pole and reminds me of home.

Several hours later, River and I set out on the path to Thami and the Nangpa Valley. We descend into the bustle of town and weave our way through its trafficked lanes, then climb back up to the brightly painted gompa on the northwest end of Namche. We pass its now quiet face, spinning prayer wheels along the way, and step out into the wide open space of the Himalayan landscape. We are glad to leave the dust and commerce and roosters and hammers and dogs and shops behind, and walk west, away from the tracks so beaten by trekkers and mountaineers, where immediately the world is a quieter place.

Groves of silver firs stand in the warm late morning sun, and the warmth seems to draw the balsam scent right out of their needles. As soon as it hits my nostrils, my head begins to spin and I am reminded of a hundred days in the north woods of Maine and the mountains of North America. The nostalgia becomes acute, and I start reflecting exuberantly on faraway landscapes. River indulges me for a while, then just laughs. Here, twenty-thousand-foot peaks mantled in snow rear up above the tops of the conifers as a constant reminder of our Himalayan geography.

The quiet calmness on the trail strikes us both, and for the first time since we've been in Asia I can actually feel my mind starting to empty and relax. We descend through the forest to the tiny villages of Dramo and Thomde, where lodges are few and most people seem concerned only with daily subsistence—preparing and manuring fields, pasting moist patties of dung against the sunny side of rock walls to dry, trafficking potatoes to adjacent villages. Unlike the villages along the Dudh Kosi, up here there are no spring-green crops breaking the brown soil. Last night's hard frost was enough to explain why. Here the sowing was just beginning.

As we walk through Thomde, two young boys emerge from behind a rock wall and begin to follow us, giggling. We turn towards them, and crouch low to meet them eye to eye. They are mischievous, with sparkling eyes and big smiles. They must be no older than five or six. Both are clad in a strange mix of out-of-date western clothes and traditional woolen garments. Their cheeks are dirty but rosy. We talk with them a little, though we don't really understand each other, and their English is no better than our Sherpa. But somehow we all find it fun and are struck by the oddity of each other. I reach into the lid of my pack and pull out a pen and a pencil. Of all the gifts a foreigner can give to Sherpa children, something to write with may be the greatest. Such things are reputedly hard to come by up here. They are pleased, and instantly begin writing on each other's hands. Just then, a small girl with enormous red cheeks crawls over to the rock wall and props herself up. She sees what is happening and reaches her brown hands out toward us. I try to explain that I don't have any more pens, and that maybe her brothers can share, but she just looks confused and spits up all over the rocks. River bursts out laughing and so do the boys. I am astonished at first, then join in the laughter. The little girl disappears behind the rock wall and crawls back to the nearby house.

We descend into the deeply incised gorge of the Bhote Kosi, where the exaggerated downcutting of the Himalayan streams becomes glaringly evident. Here, the huge debris fans of landslides are visible everywhere, spreading to wide aprons where they meet the river bottom. I look at the slopes above for signs of deforestation, and there are none. Studies in the 1970s and '80s suggested with no small amount of certainty that the Himalayan environment was in "crisis," due largely to the twin factors of increasing deforestation and agriculture associated with population growth among traditional peoples. The crisis scenario that was painted at this time was not limited to the relatively sparsely populated regions of the high Himalaya, but also affected the watershed downstream, including even the entire Ganges plain, where drastically altered sediment loads and volumes of water would change the agricultural dynamics of hundreds of millions of people. But subsequent inquiries into the "Himalayan crisis scenario" have debunked the hypothesis, asserting that the Himalayan landscape is inherently unstable, and because of its incomparable elevation and relief, what may look like ecological disaster in other parts of the world is perfectly normal here and has been for millions of years. Perhaps it is just hard to hold together a landscape that is so geomorphically active and extreme, and human effects contributing to such processes are negligible. Presently, both arguments are based largely on comparative observations and speculations, and neither argument has been substantiated by compelling data. In my own observations, I have yet to see a debris fan associated with deforestation. As for agricultural practices, the Sherpa have been here for hundreds of years, and they are not fools. They catch soil and contain it rather than cause it to slide away. If anything, the effect of so many terraced fields seems to be to keep more Himalayan soil in the Himalaya, and not send it sluicing downstream to India. The impact of grazing on the landscape, however, is profound and hard to ignore. The whole place seems cropped, mowed, cut up by hooves, and pummeled. Such is the price of milk, butter, wool, meat, and beasts of burden, without any of which traditional Sherpa life in the Himalaya would be impossible. Perhaps it is not too high a price.

We descend into the gorge at the bottom of the valley. Here, cold, blue meltwater from the snows above the valley of the Bhote Kosi have cut a slot a hundred feet directly into the underlying bedrock, and the result is a gorge of sensuously carved gneiss flutes, grooves, potholes, and arches. Beside the slot, on a flat, vertical section of stone, is a life-size portrait of a blue, round-faced Buddha sitting in the full lotus position and surrounded by cloudlike lotus flowers. Around the Buddha's crowned, top-knotted head is a great blue and red halo, and beyond it, outside the intended frame of the painting, is a small white moon and five stars. Whether it is the historical Sakyamuni Buddha or Chenresig, one of the three great Bodhisattvas, or the Guru Rimpoche, we cannot tell. The sound of the river boiling down in the slot drowns out all else, so we stand quietly staring at the blue Buddha, the cold wind flinging water droplets all around us.

I step up the dusty switchbacks that are the final approach to Thami, and I am suddenly graced by a large flock of fluttering snow pigeons doing barrel rolls in the sky just above me. Noon has passed, and the high peaks and ridges are predictably swathed in clouds. The path follows a hundred cascades and small, clear pools of cold water, running amid stands of pagodalike junipers with an understory of neatly cropped graminoids. A few buildings with walls of stone and mud come into view, scattered loosely amid wide, brown, bare fields

hemmed in by rock walls. Ahead and above, pasted to the side of a mountain like a bright, exaggerated Anasazi cliff dwelling, is the Thami gompa, home to a score of Tibetan monks. The village is quiet, but there are people about, in the lanes and in the fields, going about their everyday lives. I take shelter from the cold wind behind a low rock wall, and sort through my pack for a wool shirt and a sweater. We have walked far to get here, and our intention is to stay awhile.

Namache Bazaar

# IV

Thami on an April morning is a stunning sight. The sun illuminates the fresh snow of the high Himalaya with glaring intensity, and the high peaks and ridges seem to hum out loud in their shimmering white glow. The distant sound of rushing water is always there, though one has to stop and listen to notice it. Rectangular houses of stone and mud dot the brown, rockwall-bound fields. Morning light warms the brown soil, moistened daily now by regular afternoon rain and snow. Yaks drag primitive wooden ploughs across the fields while women in traditional dress toss mulch and manure into the furrows. Young, big-eyed calves stand against rock walls with innocent but bewildered expressions on their faces. In front of most of the nearby farmhouses, burning sprigs of juniper send a fragrant smoke into the air, offerings made by householders to God in hopes of good fortune for all.

These are precious scenes and I do not take them for granted. I know these are traditional people because they derive their sustenance from the land they live on. Although they practice animal husbandry and agriculture, they know almost nothing of the mechanization of the Industrial Revolution, or of the reductionism of Western thought. They have little if any dependence on petroleum products. The "information age" is not a phrase they would recognize. As such, they are simple folk who live rich lives, alive with all the wonder of a culture still in touch with its own mythological roots, whose deep stories are grounded in the same land they still live on. And they live hard lives, stooped over dung fires in dark, poorly ventilated houses, breathing black smoke every day. They suffer the elements and the toil of labor on a regular basis, and so are capable of enduring with a smile what most westerners would consider inhumane, intolerable, unnecessary hardship. I admire them, even envy them, and I wonder if I could ever live the way they do. To see the Sherpa in what may well be the first place they settled in the Khumbu region, still employing some of the same subsistence methods they brought with them from Tibet hundreds of years ago, is inspiring and relieving.

There are many stories about the origins of the Sherpa and how they came to live in the Khumbu region. In almost all of these stories the Sherpa came to the Khumbu via the Bhote Kosi Valley, and likely settled in the vicinity of Thami before anywhere else in the Khumbu. Most stories, of both a legendary and academic nature, point to the Nangpa La as the place where the Sherpa spilled over the high mountains and into the high valleys of the Khumbu. Their name comes from Tibetan for "east-people," *Shar* meaning "east" and *pa* or *wa* meaning "people," and from this, as well as other anthropological evidence, we gather that their original homeland was in eastern Tibet. When threatened by Mongol expansion in the late thirteenth and fourteenth centuries, it is likely that the Sherpa moved west into central Tibet, and were later pushed over the Nangpa La by subsequent conflict with Mughals during the early sixteenth century. Five hundred years ago they would have found the Nangpa La more heavily glaciated than it is today, which, interestingly, may have made the long passage over the Himalaya more straightforward than it is now.

The Sherpa were not the first to visit or even inhabit the Khumbu region, but they were the first to stay. Archaeological evidence and simple speculation alike suggest that humans visited the region over a period of thousands of years, but without the seemingly contradictory combination of specialized technologies and cultural adaptability that the Sherpa possessed, early visitors were not able to eke out a year-round living up in the steep, high slopes and valleys of the Khumbu. Eventually there was some farming in the region, but the inhabitants did not stay long enough to tell their tale. Tibetans probably pastured cattle here, accessing the Khumbu from the north via the Nangpa La, thus establishing the pass even before the Sherpa made use of it. But from what we can tell, the early Tibetans were interested in seasonal grazing grounds, not permanent settlements.

The history of human inhabitation of the Khumbu region is much like that of similar mountain regions around the globe. People naturally inhabit the lowlands first, where farming is relatively easy, growing seasons relatively long, and the variety of viable crops greater than in the adjacent mountainous areas. But eventually population increases and the resulting increase in pressure on natural resources pushes people into more marginal habitat. Necessity drives the development of the specialized technologies needed to carve a living out of such previously unusable tracts of human habitat, and people move up. Most often it is seminomadic pastoralists, already used to both marginal habitat and being on the move, who, with their inherent need for large ranges, are displaced by lowland agriculturalists and pushed up the slopes. In mountain regions of subtropical and tropical latitudes, such as the Himalaya and parts of the Andes, rudimentary agriculture is possible at exceptionally high elevations, and the only thing stopping permanent human habitation from occurring even higher is the inability of a human fetus to develop above fourteen thousand feet. The Sherpa press right up against this physiological reality, and even have seasonal habitations up to sixteen thousand feet.

The Sherpa are essentially a disjunct population of Tibetans who, over time and in relative geographic isolation, have evolved significantly enough to be considered their own culture, both linguistically and by custom. The Sherpa are a mongoloid people and bear little resemblance to their lowland Nepali neighbors in either physical appearance or culture. Their language, of Sino-Tibetan origin, still closely resembles Tibetan, and has nothing in common with the Sanskrit-based Nepali spoken by the majority of the country. In fact, all of the Indo-European-based languages of Europe, including the Romance, Germanic, and Celtic languages, are closer to Nepali by several orders of magnitude than Sherpa is to Nepali. Used to a seminomadic pastoral existence on the high, windswept Tibetan plateau, when they migrated to the south side of the Himalayan crest, the Sherpa brought with them their religion, their stories, their language, their dress, and perhaps most important, their livestock.

The Sherpa are a people of the yak. Like all traditional peoples of both high latitudes and high elevations, they are dependent on animal products, at least in part, for food, fiber, fuel, and shelter. Without such animals, human inhabitation of any but the most arable portions of the globe would be impossible. If not hunters and gatherers, the people of such places are invariably pastoralists first, and sometimes agriculturalists second, but not, by virtue of the fact that they would starve if it were so, vice versa. Among northern peoples, the Inuit, Yupik, and Aleuts of the Polar North survive on marine mammals, the Sámi on reindeer, the Ihalmuit of Canada (now extinct) and the Inupiat of Alaska on Caribou, and even the Celts and Germanic peoples on cattle. Among mountain people, the Andeans' lives were traditionally bound to llamas and alpacas, and in the Himalaya, the key to survival in the high mountains and vast plateaus was that once-wild, muscle-bound, flexible-hoofed, hairy relative of domestic cattle, the yak.

Yaks, and their female counterparts, naks, are the native bovine of the cold, semi-arid highlands of Central Asia. They were domesticated from their wild progenitors as early as the Neolithic, and although they are remarkably smaller (wild yak bulls can weigh in at a whopping twenty-two hundred pounds), variously colored, a bit less hairy, and easier to get along with, domestic yaks still retain many of the evolutionary adaptations to high, cold climates that their forebears did. They have both a short, fine, dense undercoat and a long, coarse overcoat, ideal for both insulation against the cold and for shedding snow. Their lips and tongues are adapted to cropping the short, often prostrate plants typical of alpine plant communities. Their digestive systems process food far more efficiently than most of their lowland relatives. They produce milk twice as rich in fats as that of their lowland counterparts. In many regards they are built, both morphologically and physiologically, like mountain people: short, tough, barrel-chested (to accommodate larger lungs), wide and hard-footed, with swarms of relatively small, red blood cells, hemoglobin-rich blood, and plenty of extra capillaries in the extremities.

The same adaptations that make yaks so successful at high elevations make them miserable at low elevations. Below ten thousand feet in the Himalaya, they overheat and become exhausted. They become waterlogged. They hypermetabolize. The alpine environment is relatively sterile, and with no immunities to lowland diseases, they contract livestock maladies with ease. Males, already temperamental, get worse. In short, yaks have become too specialized to be successful in a broad elevational niche. But up high, in those landscapes where they have evolved over tens of thousands of years, they are unparalleled.

Yaks are often bred with lowland cattle to produce hybrids, with the intention of producing animals that have a significant resistance to cold and altitude, but that are also more docile, longer-lived, larger, and produce greater volumes of milk. Hybrid nomenclature is complex. Male hybrids are generally called *zopkios*, and they are all sterile. But a zopkio with a yak father and a cow mother is called a *urang*, while a zopkio with a bull father and a nak mother is called a *dimzo*. Female hybrids are called *dzum*, and are fertile. Dzums are readily bred with both yaks and bulls, and there are, accordingly, second-generation hybrid names.

Until today, I have never seen a living yak. All along the footpath from Lukla to Namche, and again from Namche to Thami, I looked in vain for a pure yak, and found only the zopkios and dzums of the middle elevations. Any uncertainty as to the identity of certain bovines previously encountered is dispelled here. Before me, pulling a plough in the brown dirt of a Thami field, is an animal that could never be mistaken for a cow. Its body moves with the strength and elegance of something still close to wild. The prominent shoulder hump is reminiscent of a buffalo or a musk ox. Its long, straight coat, a fringe accentuating its powerful haunches, is both beautiful and functional. More than anything else, this animal looks like it belongs here.

The Sherpa traditionally used yaks in nearly every aspect of their lives, and still do to varying degrees today. Yaks are beasts of burden, used to working in fields and carrying loads. They are food sources, and as such they are milked, bled, and, at least historically, were slaughtered and butchered for meat. They are fiber sources, and are clipped and sheared for rope, straps, blankets, and clothing. They are fuel producers, providing the only combustible fuel source above the alpine treeline. They provide manure for fertilizer, which is so valued in the

relatively thin, nutrient-poor soils of the high Himalaya it is often carried for miles from collecting places to fields. And, seldom remembered, they are sources of building materials, providing the dung of the
dung-mud mixture used to side so many traditional houses.

Upon their arrival in the Khumbu, the subsistence methods
employed by the ancestral Sherpa shifted and evolved. Once established in the region, the Sherpa soon came to dominate the trans-
Himalayan trade of the eastern Himalaya, and enjoyed the continuous
influx of diverse goods from both the Indian subcontinent and interior Asia. By virtue of their strategic geographic position, the Sherpa
were not only able to benefit from the varied goods that passed
through their region, but also became middlemen, gaining a profit
from trade in both directions. At the outset, the Sherpa likely led a
marginal existence in the Khumbu and adjacent highlands of eastern
Nepal, and this may have led to an increasing dependence on outside
goods. The Sherpas' ability to satisfy this growing dependence, largely
due to their strategic position between India and the interior, may be
the single most important contributing factor to their shift from centuries of variations of nomadic pastoralism to becoming a people
firmly rooted in one place.

Two hundred years ago, the neatly walled fields of Thami would
never have been here. These are potato fields, and Thami is the potato
capitol of the Khumbu. I remember walking here yesterday, and
nearly all the people I saw walking to Namche were carrying a load of
potatoes on their backs. Thami potatoes reach Namche, and through
Namche they reach the far corners of the Khumbu. But potatoes are
a New World crop. When Europeans first saw them, they were
appalled at the idea of humans eating such rude food, even though
potatoes had sustained Andean peoples of South America for untold
generations. Ignorant of its superior nutritional qualities but well-
aware of its hardiness, the British and other Europeans fed potatoes
to their livestock, and to such primitive remnants of Celtic Europe as
remained, namely, the Irish and the Scottish. Eventually, the British
brought the potato to their Darjeeling tea plantations in the eastern

Himalaya, presumably as animal fodder, and possibly as specimens for their extensive botanical gardens. Once the Sherpa got hold of the same crop that had sustained their Andean counterparts for centuries, they unknowingly followed suit. Potatoes joined the existing trilogy of barley, buckwheat, and brassicas, and quickly rose to prominence as the most important food crop. With the introduction of the potato to the Himalaya, mountain people experienced an unprecedented population boom. The result has been expansion into still more marginal lands, where the robust potato alone can grow, and increased pressure on already limited resources such as fuel wood, soil, and pasturage.

Presently, the Sherpa are fertilizing their fields with mulch and manure, and beginning to plough and sow. The potatoes will grow over the long monsoon season from May into September, and harvest will come in October as the monsoon subsides. Today, every member of every Sherpa family will eat potatoes sometime throughout the day. Most of the westerners visiting the Khumbu region will do the same. We will all join a billion others in such far-flung places as Bolivia, Ireland, Siberia, and, of course, the Himalaya to give thanks to the tuber that caused a revolution among common folk the world over.

The combination of ages-old agriculture, increasing involvement in trans-Himalayan trade, and the introduction of the potato all contributed to a growing dependence on place and field. But aspects of nomadic pastoralism persist in Sherpa subsistence methods, even to this day. While valleys are cultivated and speckled with villages, the surrounding slopes and ridges are grazed by livestock. In the spring, the yaks and naks are brought down from upland pastures to plow the valley fields. The yaks spend their off time in an outer grazing perimeter, while the naks remain in the villages for milking. In summer, yaks and naks alike, as well as most family members, move up to seasonal habitations at higher altitudes, and livestock spends the wet summer months grazing on the verdant green summer alpine vegetation. Family members tend animals, in some places cultivate high-elevation grains, and collect dung to be used as fertilizer. Only the elderly remain in the villages, where they tend the summer's crops, occasionally weeding the fields for greens for the pot. By autumn, the grain crops are harvested and the family and livestock move back down to the valleys to harvest lowland crops. As winter presses on, the yaks, accompanied by a few hardy herdsmen, are brought back up high to forage for vegetable matter amid windblown snow and frozen ground. The naks, meanwhile, enjoy less rugged accommodations in the valley villages, and are milked dry through the winter. Western academics have termed this rare traditional subsistence method *transhumance*.

The frequent seasonal splitting up of families led to the unusual marital practice of polyandry. Brothers often share a wife. First sons typically marry, and if they are lucky or work exceptionally hard, they inherit or establish a herd and property. Second sons are typically sent to the local monastery. Third sons are either married to the first son's wife or, in some cases, are adopted by their older brother. Admirably, traditional Sherpa society is very egalitarian. Both men and women own property, and though most forms of work are typically done by one gender or the other, such social norms are loose, and it is not unusual to see men cooking and taking care of children while women hoist heavy pack baskets and tend livestock in the high country.

Before the mid-twentieth century, transhumance was practiced by nearly every Sherpa family in the Khumbu and adjacent areas. Today it is practiced by less than 10 percent of the population. The vast majority of Sherpas have adapted to yet another influx from far away: Western tourism. What began as a few young men employed as guides and porters has turned into a flood of jobs, including mountaineering guides, cooks, porters, lodge managers, and shop owners. Ever opportunistic, steadfast as a people, but amazingly adaptable to meet the needs of changing times, the Sherpas' ability to turn what could potentially be a cultural downfall into a cultural advantage has made them one of the richest tribes in Nepal. Today, one must travel far, and away from the names of well-known places, to see the last remnants of traditional transhumance still in practice. The valley of the Bhoti Kosi is such a place, though its future is perhaps in question by my very presence here.

Clouds obscure the mountains, and by noon, Thami appears to be floating in a gray-white void. I pile on clothes, but the wetness of the cold air sends a chill into me that stymies my best efforts at warmth. I stand, leaning against a rock wall, looking out over the brown fields. Snow begins to fall in large, heavy flakes. At first they fall gracefully, like a thousand white feathers floating to the ground. But within minutes they thicken into a whirlwind. They melt into the warm ground for a while, then stay, decorating the dirt with a thin, frosty white veneer. I wonder if the snow is exceptionally late this year, and if the sowing will prove foolhardy. But before my eyes, ten Sherpa women throw mulch in furrows, undaunted by the weather. As I stand to leave, the rock wall I have been leaning against gives way. As the rocks clatter to the soil, the women all look up, instantly, and there is a second of silence. Then one of them starts yelling at me crossly in Sherpa. I am mortified by what I have done. But her tone quickly changes, and soon all ten women are laughing out loud, and so am I.

Thami

A yak

# V

The master of this house, the Cho Oyo Lodge, is a good man. He is pious, and though a householder, he keeps his Buddhist spirituality at the center of his life. As I sit and eat he sings a song, and it is strangely reminiscent of Native American songs I have heard. While singing, he walks over to the chorten, or small altar, at the end of the room, where a dozen small, ornate water bowls are arranged on either side of a central candle, next to which is a tiny glass of tea on a pedestal. On the wall behind this assortment of religious objects is a picture of the Dalai Lama, pictures of other lamas, and other objects of Tibetan origin that remain mysterious to me. He methodically empties each of the bowls (which I watched him fill this morning) into a bucket, then stacks the bowls in two upside-down piles on either side of the central candle. He completes this ritual, still singing, by removing the tiny glass of tea from the chorten, opening a window, and casting the tea out the window. He does this with a few unfamiliar words, said loudly and obviously intentionally. I ask him what the tea is for but he understands almost no English. I try again, this time in a combination of mispronounced Sherpa words and comical gestures. He smiles widely and his eyes disappear into slits. Either he understands me or just thinks I'm ridiculous. He pauses for a second as if to remember something, looks at me clear-eyed, and says "for God."

He spends the majority of his time in the small, dark room that is the family's kitchen, stooped over the smoky dung cooking fire cooking meals for his few guests and, after everyone is taken care of, his family. He is almost always around this building. Between meals, he stands outside in the late morning sun, burning juniper offerings, smiling at others, collecting water, fixing things. I am in awe of his apparent humility.

In the night I lie awake. A respiratory ailment I contracted in Kathmandu has gotten worse, and with this, in combination with the altitude, I am spent. But the thin air makes for restless sleep, and after hours of trying to rest and waking up short of breath, I finally decide to stop trying. Once I stop trying I fall asleep. Blue Buddhas visit me in my dreams.

In the morning sun I walk through the village and up onto the old morainal hill that binds Thami to the north. A trail follows the spine of the hill and leads up half a mile and five hundred feet to the Thami gompa, where two dozen monks live out the traditional monastic life to the best of their ability. I have studied religion academically and Buddhism in particular, and even spent time with some important Rimpoches, but I have never seen a traditional gompa rooted in the place that birthed the Buddha.

As I walk up to the Thami gompa, backdropped by front-lit, snow-bound Himalayan mountains, the monastery seems tiny and insignificant, and I think about the relationship between religions and the landscapes that birth them. The monotheistic traditions of Judaism, Christianity, and Islam all came out of the desert, and the clash of cultures that has characterized the Middle East for millennia. The starkness of the desert inspired spiritual asceticism from early times, and it could be that the sense of solitude so unavoidable in the desert contributed to the idea of one God. The conflict between different cultures furthered the notion of monotheism and made God personal, gave him paternal characteristics, and had him forging exclusive compacts with his chosen people. Among the Jews this God was called Yahweh. The Muslims called him Allah. The Christians seem to have forgotten his Hebrew name and simply call him God.

In Asia, things went differently. In the steaming jungles and dusty plains of the Indian subcontinent, Dravidian polytheism was the norm until Aryan invaders of Indo-European origin spilled into the region over the Khyber Pass. These new insurgents brought with them their language, religion, and social systems. But however rigid the structure of the Vedas might have been, the imported belief system quickly adapted to the new environment, and the diversity of the subtropical landscape inspired as vast an array of deities and demigods as any religious tradition has ever known. But looming above the subcontinent and binding it to the north was the snow-covered landscape of the mighty Himalaya. What influence did the high mountains have on the development of Hinduism?

Among scores of gods and goddesses, the Hindu tradition recognizes a trinity of deities who together represent the ever-changing nature of the world. The first member of this trinity is *Brahma*, the creator, who brings things into being and breathes life into the world. The second member is *Vishnu*, the preserver, who sustains life and being through time. The third member is *Siva*, the destroyer, who makes wreckage of being and life, and reduces the efforts of Brahma and Vishnu to utter ruin. Thus the slate is clear for Brahma to come and make the world anew and Vishnu to hold its course. But without Siva, the evolution of the world would stop, and there would be no continued renewal. I believe Brahma has his origins in the subtropical jungles, the primeval gardens of Eden. Vishnu has his origins in the plains, those fertile lands that precariously sustain human life over time. Siva comes from the mountains, the makers of storms and floods, a landscape constantly being destroyed and reborn. It seems appropriate that Siva is the most widely worshipped god of the trinity the further one goes north in India, right up into the long shadow of the Himalaya.

If Judaism, Christianity, and Islam were born of the desert and Hinduism was born of the diverse landscapes of the Indian subcontinent, Buddhism, especially that brand of Buddhism known as Tibetan, or *Vajrayana*, was born of the mountains. Here the high Himalaya force humility on the human spirit, and life closer to death leads to a greater understanding, thus compassion, for all life. Here, the Himalaya rise up from India just as Buddhism rose up from Hinduism.

Buddhism evolved from Hinduism much like Christianity evolved from Judaism. Both Buddhism and Christianity are the by-product of the effort of a religious reformer who challenged the stagnant traditions of his own religion. In the case of Christianity the reformer was Jesus of Nazareth. In the case of Buddhism the reformer was Siddhartha Gautama, known variously as Gautama Buddha, the Buddha, the historical Buddha, and among Tibetan Buddhists, Sakyamuni Buddha.

History tells us that the Buddha was born around six hundred

BCE, and was raised at the foot of the Himalaya, in the Terai region of what is now southern Nepal. He was born a prince, the son of rich and powerful king. Upon his birth, his father summoned fortune tellers to forecast the fate of Siddhartha, the young prince. It was foretold that if he remained in the world he would become a great warrior-king and unite India under one dominion. If he forsook the world he would become a great religious reformer. The Buddha's father, appalled at the idea of his son becoming a religious fanatic, went to great pains to insure that young Siddhartha remained in the world and never experienced any of the hardships in life. He went so far as to shield the young prince from sights that might raise spiritual questions in his mind. Thus, young Siddhartha grew up in a rich, beautiful world devoid of sickness, pain, or death.

One day, as legend tells, it was a day of festival, and Siddhartha rode out from his palace as he had done so many times before, his loyal friend and servant Govinda by his side. But his father had been negligent, and had forgotten to clear the streets of those things which might inspire deeper questions in the prince's developing mind. As Siddhartha and Govinda rode forth, an old, bent man, teeth yellowed and broken, hair spare and gray, hobbled into the street leaning on a crooked cane. Siddhartha stopped, wide-eyed, and watched the old man go by. Further down the road he saw a man struck with disease, his body disfigured, his flesh ruptured by legions. Then he saw a corpse, an empty shell with all life gone out of it, lying in a ditch. Finally, as if in answer to the questions that must have been stirring in the prince's mind, he saw a monk, robed in ochre, head clean-shaven, carrying nothing but a small bowl.

Having seen the four passing sights, Siddhartha's life was changed. He could not return to the pleasures of the world, and instead he contemplated renouncing it. Finally, in his twenty-ninth year, he kissed his sleeping wife and child goodbye, went out into the night, shaved his head, stripped off his clothes, and walked into the forest.

Siddhartha lived in the forest for six years. He sought out the wisest Hindu yogis and from them learned the deep wisdom of raja yoga. But eventually he exceeded their wisdom and moved on. Next, he joined a group of extreme ascetics, and with them he endured every bodily hardship imaginable. But in living so close to death and still not attaining enlightenment, he realized the futility of asceticism. Through the former experience he learned mystic concentration and focus of thought. Through the latter practices he learned that the way to enlightenment was neither the path of extreme indulgence nor that of extreme asceticism, but rather the middle way between the two.

These two great lessons prepared Siddhartha for his third and final step toward enlightenment. Perhaps sensing that enlightenment was imminent, he sat down one evening beneath a banyan tree and vowed to not rise until he was awake. Then darkness fell. Mara, the Evil One, descended upon Siddhartha with fury. He turned into seductive goddesses and paraded around Siddhartha, offering an infinity of sensual pleasures if only Siddhartha would relent. He bore the guise of Death and summoned a thousand armies to assault the serene Siddhartha with weapons, fire, flood, and darkness. But the maidens went unbidden and weapons, fire, flood, and darkness turned to lotus petals, sweet rivers, and golden light. Finally, Mara challenged Siddhartha directly, asking by what right Siddhartha sought enlightenment. At this Siddhartha reached down with his right hand and touched the earth on which he sat, and the earth responded with a resounding roar: "I bear you witness!"

Siddhartha remained beneath the tree through the night. Then, just as the darkest hour relented to the first hint of dawn, he penetrated the deepest recess of the universe, and found the emptiness of the void. In the depths of this emptiness he realized the true nature of being. This was the supreme moment, the Great Awakening, the first nirvana, and the entire universe resounded in themes of awe, wonder, and cosmic rejoicing. Siddhartha remained there in a state of profound bliss for seven days. When finally he tried to rise he was engulfed in bliss a second time, this time for forty-two days. When finally he opened his eyes to the world he had become the *Buddha*, the "Awakened One."

Mara was not quite done with Siddhartha. As the Buddha opened his heavy eyelids to the bright light of the word, Mara was there waiting for him with one last temptation. He was not only threatened by the Buddha's singular enlightenment but terrified at the possibility of such awakening spreading to other beings. Recognizing that the Buddha had seen through the illusion of the material world, Mara posed a question that struck at the most deeply seated desire of every spiritual seeker. In the poetic words of Huston Smith, "Who could be expected to understand truth as profound as that the Buddha had laid hold of? How could speech-defying revelation be translated into words? How could visions that shattered definition be rendered in terms of yes and no? In short, how could he show what can only be found, teach what can only be learned? Why bother to play the idiot before a jury of uncomprehending eyes? Why try to argue the glance of spirit to masses still caught in the whirligig of passion? Why not commit the whole hot world to the devil, be done with the body forever, and slip at once into the cool haven of perpetual nirvana?" The Buddha wavered. Mara's arguments seemed so seductively convincing. After all, Siddhartha had worked so hard for this moment of supreme enlightenment. . . .

In the end the Buddha prevailed, saying simply that there would be some who would understand. Mara and his legions wailed and fled in terror, defeated. The Buddha went back out into the world, and for nearly half a century he wandered the Indian subcontinent teaching the masses. His message was and still is simple yet profound. At its core are the four noble truths. The first, *all life is dukkha. Dukkha* is the condition of the material world that inherently involves suffering. The second, *dukkha is caused by attachment and desire.* Third, *attachment and desire can be overcome.* Fourth, *attachment and desire can be overcome by following the eightfold path.* The Buddha also laid out the eightfold path with stunning simplicity and clarity. It goes something like this: *right understanding, right thought, right speech, right action, right livelihood, right effort, right mindfulness, right concentration.* The Buddha traveled far and wide expounding this great message to the world. He never wrote anything down, but his students diligently recorded many of his most profound teachings. They are the Diamond Sutra, the Sutra on Totality, the Heart Sutra, among dozens of other sutras, parables, and assorted teachings.

India embraced the Buddha. His personality was magnetic, his ambiance clarifying, his intellect sharp, and his heart filled with overflowing compassion. He founded some of the oldest monastic traditions in the world, and during his lifetime his monks numbered in the thousands. His teachings reached hundreds of thousands, and eventually millions, challenging the doctrines of orthodox Hinduism with the notion that each and every being should seek its own enlightenment, free of the corruptions of authority, ritual, philosophical speculation, tradition, and sovereignty that had long since plagued and disabled the functionality of Hinduism. It was social reform, religious rejuvenation, spiritual revolution.

As I approach the gompa, the simplicity of the Buddha's doctrine seems far away, a dim light set back in a dark hall of time. Several stone buildings, a few of them plastered with mud and dung, are built right into the rocky slope. They are unassuming, blend in soothingly with the surrounding landscape, and bear none of the ornate trappings that I have come to associate with established religions. They are subsidiary buildings, however, centered around a main building that seems to have sprung out of a completely different line of thinking. The gompa itself is painted a bright ochre hue that contrasts sharply with the landscape around it. Each and every window is framed in an angular black field, within which are brightly painted blue and white window jams, and mustard-yellow frames. Inside each frame are intricate paintings that obscure the view inside, or meticulously embroidered curtains. Above the heavily curtained doorway is a decorative mantelpiece that looks to be either the eight-spoked wheel of Dharma or an impression of a lotus flower, or perhaps both.

I step inside the doorway, spreading apart thick, black, cloaklike curtains. Inside it is dark, silent. As my eyes adjust to the musty, dim light I can make out an immense prayer wheel, elaborately painted with prayers and scenes from the ancient Tibetan tradition. I grab a loop of wool that hangs down from the bottom of the cylinder, and give it a pull. The wheel creaks as it turns, and I think to myself that these creaks in the dark and dust are the sounds of a religion grown old and in disrepair, in need of more refreshment than the façade of another bright paint job. Beside the prayer wheel is a closed, locked door with a donation box next to it. The box is stuffed with rupee notes. The place is empty, the monks either gone or locked up inside, sitting on benches and clinging to prayer beads, humming ancient mantras over and over and over again. This place is fascinating, totally foreign, and incomprehensible to me. I feel reverent, even respectful, but not spiritual in the least, as if the heavy overlay of cultural décor here has somehow obscured the truth of the Buddha's teachings. I am reminded that while the central teachings of any religion may be culturally transferable, the culture, traditions, ritual, and practice that surround the religion often are not. I leave the inside of the gompa confused, and step back out into the light of the day, into the cold, crystal-clear mountain air.

Tibetan Buddhism is rife with the authority, ritual, philosophical speculation, tradition, and sovereignty that the Buddha eschewed, and the complexities and nuances of its practice are a far cry from the simple truths of the man for whom the religion is named. The changes reflect a combination of a long history of divergence in Buddhist thought and the influence of landscape, and thus culture, on religion.

Buddha was the glue that held Buddhism together, and soon after his death Buddhism began to experience divergence. The question that was most disagreed upon was the nature of being and its relation to enlightenment. The earliest recorded teachings of the Buddha seemed to indicate that man was an individual, alone in the universe, and the highest aspiration of the individual was wisdom, enlightenment, and attainment of nirvana. But the actual life of the Buddha suggested rather that all beings were part of an inextricably connected community, that enlightenment for one was ultimately contingent on the enlightenment of all beings, and the highest aspiration was wisdom that leads to compassion, and compassion that leads enlightened beings to renounce nirvana, as did the Buddha, in favor of dedicating their lives to helping others.

This fundamental difference in interpretation of the Buddha's life and teachings led to the development of two distinct branches of Buddhism. *Theravada* Buddhism (the Way of the Elders) followed the principles of individual enlightenment, the path of wisdom, and made monastic life its focus. The ideal of the Theravada tradition became the *Arhat*, the ideal disciple, who makes the attainment of nirvana their sole focus in life. *Mahayana* Buddhism (the Great Vehicle) followed the principles of community enlightenment, the path of compassion, and made laypeople its focus. The ideal of the Mahayana tradition became the *Bodhisattva*, or saint, who, at the threshold of nirvana, renounces its bliss in favor of remaining in the world to help others along the path.

Over the centuries, the Theravada tradition became more exclusive and its influence waned, while the Mahayana tradition became more inclusive and waxed. While the Way of the Elders strictly adhered to the Buddha's abhorrence of authority, ritual, philosophical speculation, tradition, and sovereignty, the Great Vehicle commenced with a flowering of hierarchical priesthoods, religious rites, divine cosmologies, elaborate traditions, and eventually transformed the once very human Buddha into a savior-god. Hinduism eventually won the day in India, and subsumed the Buddha and his teachings into its own long list of deities, demigods, and saints. The Theravada tradition was established in much of Southeast Asia, and as a conservative tradition its evolution all but stopped. The Mahayana tradition, inherently more accepting of cultural influences, was imported to and thrived in China, Japan, Korea, Mongolia, and Tibet, where it mixed with existing religious and philosophical traditions and further diverged. In China, Japan, Korea, and perhaps Mongolia, Buddhism met Taoism and became Ch'an, or Zen. In Tibet, Buddhism met the native, polytheistic B'on tradition, rich in ritual, tradition, prophesy, legend, and cultural color. From this meeting *Vajrayana*, the Diamond Vehicle, Tibetan Buddhism, was born.

I sit quietly in the warm sun on the patio of the Thami gompa, and watch closely as the first clouds of late morning begin to shroud the surrounding mountains. In early morning their front-lit lines were so sharp, so clear, with a crystalline definition. By midmorning they were no less bright, but less sharply defined as the world around them brightened, too. Now, late morning, they show the first signs of becoming overwhelmed. By noon they will have become obscure, glimpsed only occasionally through shifting, swirling clouds. By afternoon, the very mountains that caused the weather will seem like distant memories, and without having seen them earlier, one would doubt they ever existed. Rain will fall on the lowlands, and snow in the highlands, saturate the world, and bring to it the songs of water and life. It occurs to me, then, that pure religion, like pure light, makes clear the nature of the world. But culture, like weather, nourishes it and gives it its story.

Thami gompa

# VI

I return to Thami by the moraine path and look out over the fields and houses of the small village. Outside one building, a few dozen children, all dressed neatly in navy blue, are gathered in a circle. Inside the circle is a small man, and as the children sing a song he dances and claps in the midst of them. When they stop singing he points to one of the children and they all start laughing. This goes on a few more times as I stand and watch, trying to figure out what they are doing. It strikes me that the sounds of children are the sounds of a living village, a people in place with a future.

In the afternoon, having spent a few hours indoors reading, I venture outside to breathe some fresh air. The mountains are gone and the village floats in an impenetrable pewter-gray swath of cloud. Scattered snow flurries twirl their way downward from the opaque sky and betray the recklessness of the wind. The villagers are leaving the brown dirt and debris piles of the potato fields and turning in for the day.

As I turn to go back inside, a small, thin, brown man bundled in several layers of clothing approaches me. He introduces himself as Sibi, and noticing that he is not Sherpa, but Nepali, I ask him how he spells his name in English. "C.B.," he says, with a toothy, sparkly-eyed grin. He tells me that he moved up to Thami recently, and he teaches at the small school in the village. At once I recognize him as the schoolteacher in the circle of children I saw this morning, and that vision of him singing and dancing and clapping makes me at ease with C.B. Since he's been here he has been trying to start an interfaith meditation center, but with the religious conservatism typical of rural regions all over the globe, he has had little luck in Thami. The traditional Tibetan Buddhists of the Khumbu have little tolerance for the liberal religious ideas of the cosmopolitan lowlands. While the lowland Hindus categorize Buddhists as being somewhere near untouchables in the caste system, the highland Buddhists regard Hindus with skepticism and sometimes evident disdain. Here, C.B. is an obvious minority. Even though there are a hundred Sherpas here in Thami whose religious practices are not unlike those of the Hindus, every day C.B. meditates alone.

Somehow we get to talking about spiritual matters, and C.B.'s third-world openness is contagious. Soon we are sharing the particularities of our spiritual belief systems, and C.B. invites me and anyone else at the lodge who is willing to join him for a meditation session later in the afternoon. I accept instantly.

A few hours later, River, an American named Jeff we befriended some days ago, and I walk through the snow flurries to the small, dilapidated mud and dung house that is C.B.'s meditation center. The house, like all traditional Sherpa houses, has two stories, the lower level open with a dirt floor, previously a byre for livestock, now primarily used for storage. A steep wooden stairway leads up through a hole in the floor to the second level, which is a single open room with a creaky wood-board floor and a few drafty windows. Strewn about the floor are a few woolen mats, sitting pillows, and blankets. In the corner, of a conspicuously different aesthetic, is a large, black, plastic, battery-operated boom-box of which C.B. seems immensely proud. There is no heat in the room. We all huddle on the floor and wrap thick blankets around ourselves.

C.B. explains a few different meditation exercises and after some discussion we agree on a long sitting session with some background music of C.B.'s choice, followed by a shorter session listening to a cassette of some lessons by C.B.'s spiritual teacher (the latter we agree on to appease C.B., who was anxious to share these particular teachings with us). C.B. warms up with some kundalini yoga exercises that are vaguely familiar to me, and he practices them with the utmost sincerity. Then we all wiggle around on the cold floor and get comfortable, shifting blankets and scooting cushions under and around us.

The music is loud and all-pervasive, but it is soothing, an interesting mix of traditional Indian instruments and modern synthesized sounds. On any other day I would chuckle at such music, but today, somehow, I just let myself go with it. The music has an elemental feel to it, something between water and air, and I alternately feel as if I am melting and floating. Within minutes I become absolutely still and comfortable. I count my breath until I lose count and then start over again a dozen times. Then I lose track and let my mind go. A long time passes and my mind is quiet. Such moments are rare for me, and it feels as if I am suspended in space and time in the universe. Eventually, the sanctity of the timeless moment is broken, and I visualize my own deep, dark shadow as a tall, black, slender creature with the head of a horned goat. I have seen this creature before and struggled with him. This time we sit facing one another, eyes closed in deep concentration. At first there is tension, as if our wills are fighting, but eventually we both open our eyes and look up, surprised to see each other still there. Then we shake hands and laugh out loud together about how scared we've always been of one another. We are sitting by a river, and I send him down the river on a leaf. Then I send my dad down the river on leaf. Then I send my girlfriend down the river on a leaf. Then I send down God. Then a pillar of light comes down through my head and out my butt and I am suspended in emptiness.

Eventually these deeper meditations fade to thoughts and I find myself planning things and organizing in my head. Once I am aware of this I open my eyes. It is like waking up from a profoundly restful sleep. Next to me, Jeff, having abandoned a sitting meditation, shuffles around noisily doing hatha yoga like a western gymnast. Across the room, River sits like a stoic in traditional Renzi Zen style. Next to her C.B. is still, and wrapped in so many blankets and hats that his small brown face is the only part of him that is recognizable. Snow falls heavily outside, the crowns of the junipers that surround the fields of Thami completely covered, the ground white, and the air thick with Himalayan snowflakes. On the other side of the world, men in turbans kneel on carpets in the sand. Congregations stand together and recite the Lord's Prayer in unison. Here, a score of monks sit in silence up the hill. A pious man empties the water bowls of his chorten. Three westerners sit in a cold room with a small Nepali man and meditate to music that comes from a plastic radio. I am reminded that each of us has our own experience of truth, our own stories, our own traditions, but somehow the happening truth itself is the same. There are a thousand ways to get to the top of the mountain, but there is only one mountain.

# VII

Morning. Bleary-eyed and misty-headed from too much chang the night before, we say goodbye to the village of Thami and walk into the snow and clouds of the valley of the Bhoti Kosi. It feels good to walk, and with a belly full of oily eggs, Tibetan bread, yak butter, and tea, I feel strong and ready to engage the day. The valley bears evidence of intense past glaciation, and soon after walking up and over the huge moraines that bind Thami, the broad expanse, flat bottom, and steep sides of the upper Bhoti Kosi watershed come into view. I smile at the sight, as so many memories of glaciated mountain landscapes flood my brain. But though the valley shows the signs of glaciers past, the remains of the once gargantuan valley glaciers of the Khumbu region are still far to the north.

We walk behind a yak train for some time, and I follow the animals closely, listening to their tinkling bells and watching the local herdsman at work. There are no hybrids in the group, and as I look out on the hillsides at the free-ranging cattle, they all bear the telltale shoulder hump and long, skirt-like hair of purebred yaks and naks. All of the animals, even those grazing far off on the snow-covered slopes, are outfitted with traditional handwoven wool saddle blankets and harnesses, tassels, and hand-stamped bells. We are too high for hybrids, and out of range of the influence of plastic. We are far away from the trekking centers of the adjacent valleys. There are no trail signs or markers. The few people we see are herding yaks or moving goods from village to village. No one here speaks a stitch of English. A tiny, weathered, ancient herdsman gives us a bewildered toothless grin as we walk by.

After several miles we cross the Bhoti Kosi and begin a long, ascending traverse of the east side of the valley. As I look up the steep slope ahead I keep trying to imagine where the tiny village of Lugare could possibly be. The land just looks too steep. Somehow we get off the track, and the path we are on starts to lead us further upvalley, ultimately toward the Nangpa La. A local herdswoman sees us and comes to our aid. With short, abrupt gestures she indicates that the trail is higher up on the east slope. Immediately we turn and start making our way up the slippery, snow-covered alpine heaths towards the trail to Lugare, many hundreds of feet above. I slip a step down for every two steps up, struggling and gasping for more air on the slick slope. I think to myself how comical we must look to any watching herders, two westerners loaded down with packs, off route, climbing and sliding down the slope in apparent slow motion. Finally, I reach a small bench in the otherwise monotonous slopes, and as the few scattered stone buildings of the village come into view, I crumple onto a large boulder and rest.

Lugare, at over 14,700 feet, is a seasonal village. Its few buildings are rough, and scattered widely amongst rock-walled livestock pens over the long, narrow bench. The village has no center, no shops, no mani stones, no chortens, no trees, and barely any people. None of the signs of year-round life are here. At this elevation, humans can't reproduce. The added stress on the human body of the lower levels of atmospheric oxygen at this altitude means that physical maintenance is enough, and humans can't afford to expend the extra energy necessary for reproduction. Simply put, females can't ovulate up here.

A year or so ago we would have been stopped by armed guards downvalley and we would never have been able to come close to Lugare. In the short amount of time since the guard has been relaxd, the resourceful Sherpas of the valley have managed to helicopter in the parts for a small prefabricated lodge. The few people here are busy putting the finishing touches on the lodge, and as River and I arrive in Lugare they are having their break, taking shelter from the snow and wind in a large canvas tent, swilling steaming cups of milk tea and smoking hand-rolled cigarettes. Upslope, the afternoon mist swirls around the high ridges, and ephemeral splotches of blue sky bless the day. Yaks crop the alpine meadows immediately surrounding the village. I fill my water bottle from a small meltwater runnel off the roof of the lodge. We turn in early to sip tea, read, sleep, and acclimatize.

I sleep fitfully, lingering in a haze of half sleep between gasps for breath. Part of me wants to stay the approach of dawn so I can rest, and part of me can't wait for it to come so the long, sleepless night will be over.

We rise before 5:00 A.M., and fifteen minutes later we step outside. The bright silver orb of the waning gibbous moon sits aglow above the serrate spine of the Himalaya to the west, its light mixing with the first pale light of day. The sun has not yet risen and the snow-covered landscape is gray-blue in the predawn light. The sky is varying shades of purple and lavender and seems to emit a low, barely perceptible hum. We meet the quiet of the world with silence of our own, and without a word we strap on our packs and set out into the cold air. The ground is frozen solid, and the thin dusting of snow covering the old pastures feels like Styrofoam beneath the purposeful movement of our feet.

Hidden somewhere along the high ridges to our east is the Renjo La, and through it we hope to gain safe passage to the valley of the Dudh Kosi. The pass is at eighteen thousand feet, and never having been that high, I am anxious about the altitude. In addition to the altitude and relative obscurity of the pass, the new snow will mean challenging route-finding once we leave the valley. But even though I am anxious, I am equally excited. Today will be our first day in the snow-covered, wind-scoured, lake-studded, rock-choked, bright world of the high Himalaya.

As we begin to ascend, River sets a mechanical pace, slow and steady, with breaths and steps in synch, and we transition into a silent, trance-like state of upward movement. Cairns, precarious towers of rock, stacked impeccably to ten feet high, or crumbled to dislocated piles, dot the slope, and we follow them, connecting the dots. Who made these? How long have they been here? There is something primal about the simple act of people stacking rocks, something that has been going on for hundreds or thousands or millions of years. In places, the snow obscures the cairns, and sometimes we have to pause and look. But mostly we just keep moving upward and the cairns appear.

We top an old, soil-covered moraine and the grade lessens conspicuously. I believe from consulting our map that we have climbed about two thousand vertical feet. A wide glacial basin comes into view, shadowed by the high ridge to the east. Shafts of sunlight penetrate the serrated gaps in the ridge and illuminate the peaks to the west. But look as we may, neither of us can see a breach in the ridge, and the Renjo La seems like lowland fiction. Over an hour has passed since we set out. We know from the view ahead that many more are likely to pass before we find our way to the valley of the Dudh Kosi.

The morning has been silent except for the sound of our own footsteps on cold snow and the occasional explosive songs of little brown winter wrens. Winter wrens here, in the Himalaya, at over sixteen thousand feet! We are comforted by the presence of this little forest bird from home, and are reminded of summer days in the conifer woods of the mountains of the American West. But we are far from there, though the liquid notes of the wren are much the same.

We move quickly across the basin, and huge Tibetan snow cocks cluck and scamper across what has become a whole mess of concentric moraines. We catch fleeting glimpses of these plump birds and I am reminded of the white-tailed ptarmigan of the alpine Sierra Nevada, blue grouse of the western conifer forests, and ruffed grouse of the North Woods. But the Tibetan snow cock dwarfs them all and is twice as big as the largest grouse I have seen. They match the scale of the immense mountains that are their home.

I lose the cairns for a while but soon spot a few large ones shining in a patch of sun just northwest, at the foot of a prominent ridge that rises to the main crest. We make for the illuminated cairns, and once there we enjoy our first rest, some food, and water. The sun feels good, but once in it I sit with my back to it, smear sunscreen all over my crimson face, grease up the backs of my hands and the sides of my neck, cover my neck and brow with bandanas, and girth-hitch my noseguard to my sunglasses. River does the same. Here, the mountains truly surround us and the undeniable feeling of the high Himalaya permeates us deeply.

After a good, long rest, we hoist our packs and begin looking for cairns. But there is only one, and beyond it just snow, rocks, and pregreen alpine vegetation. The right side of the ridge looks too steep to be the route, so I follow the left side into a wide, dry basin at the head of which are two distinct peaks with an obvious gap between them. Unsure of the route, I stop and wait for River. The landmarks match up with the map almost perfectly, but something seems off and we can both sense it. After some deliberation, River goes back to the top of the ridge to have a look around to the right. When she returns, she reports that she has found another, identical basin just to the northeast of us. We dally for almost an hour, trying to make the decision about which way to go. Without a compass, our only tools for navigation are the map, the land, and the sun. Finally, after careful observation of the position of the sun relative to the basin, I discern that the sun is too far to the west. I think River was onto this the whole time, but she was either too polite or riddled with self-doubt to say anything. But what about the ten-foot-high cairns that glistened in the sunlight? If the right-hand basin was the correct route, what were these huge cairns doing way over here, half a mile west of the true route? My ego is bruised and my error has cost us a precious hour or more, not to mention the extra energy expended on the additional travel.

We traverse the ridge to its right side, staying high so as not to lose the ground we have already gained. The slope has fortunately gotten the first sun of the day, and yesterday's snow has melted off, otherwise a traverse would be impossible. We skirt the forty-degree grade, terraced by grazing yaks. We continue over talus and scree, aiming for a cliff band that connects to the top of the glacial riser nestled in what we hope is the correct basin. Atop the riser is a Little Ice Age moraine composed of freshly hewn, angular rocks. Bound by the moraine is a large tarn, partially frozen, over a quarter mile across. It is not on our map. Neither is the prominent ridge separating this basin from the one we mistook for the route. I look up and around at the surrounding mountain wall, and etched into it is a seemingly insignificant col, a subtle snow-covered, talus-strewn, rocky pass. It is the Renjo La. It must be. It is past ten o'clock. In less than two hours this whole place will be socked in.

As soon as I leave the tread I become aware of the altitude. The labor of being above sixteen thousand feet hits me suddenly, like a bag of wet sand. My breath becomes heavy, too close to desperate, and my legs seem like they are made of lead. I convince myself that this is normal, to just keep moving, and I do. Soon I begin to traverse steep, snow-covered talus, and it becomes impossible to know where to step. After a few slips into leg-breaking holes I stop, unstrap my ice ax, pull on my gloves, and start probing. Looking ahead, I realize this is what it is going to be like for the next thirteen hundred vertical feet up.

I focus my concentration on the ground in front of me and keep moving, taking one step on each inhale, one step on each exhale, locking into a slow, mechanical pace. Time slows down, and even the act of resting on my ice ax for ten breaths becomes work. I feel exhausted, and I know the thin air is playing tricks on me. The Himalaya are forcefully bowing me down. I am no Olympian athlete, but I know from years of experience that if there is one kind of terrain I excel in it is steep, rough, uphill terrain just like this. But now I am struggling, and every ten steps seem like a tremendous battle of both bodily and mental strength. With every step I reconvince myself that I have it in me to make it up this pass.

Approaching eighteen thousand feet, the lightheadedness of being in the high mountains turns into the dull throb of oxygen deprivation. Here, the barometric pressure is half of what it would be at sea level. Not only are there fewer oxygen molecules in a given volume of air, but the body's ability to metabolize oxygen is significantly inhibited by the decrease in the weight of the air column above. Lungs cannot contract as effectively due to the lack of pressure, and lose efficiency. Brains swell, and their demand for oxygen increases. At eighteen thousand feet, the amount of oxygen the body is able to metabolize is half that at sea level. Overall, bodily tissues become starved of their mainstay, and increased oxygen depletion leads to decreased tissue functioning. The medical term for this condition is *hypoxia*. It is a fancy word for slow death.

Even at four thousand feet above sea level, there is a measurable decrease in aerobic performance among humans, and although few would call such an elevation high altitude, in the strictest sense this is where oxygen deprivation begins. Medical sources consider eight thousand to twelve thousand feet to be high altitude. At these elevations, most humans experience varying degrees of lightheadedness, fatigue, and shortness of breath, and although life-threatening complications are not common, they can occur. From twelve thousand to eighteen thousand feet is considered very high altitude. Without specialized physiological adaptations to such environments, the human body requires a period of acclimatization to physiologically adjust to the decreased oxygen and air pressure. Above eighteen thousand feet, where available oxygen is less than 50 percent than at sea level, is considered extreme high altitude. Humans are incapable of fully acclimatizing to such altitudes, and the body begins to deteriorate. At eighteen thousand feet such deterioration is slow and may take many months. At twenty-four thousand feet it may take only weeks, sometimes just days. At thirty thousand feet it is only a matter of days or hours.

Acclimatization occurs as the body adapts to changes in air pressure and available oxygen. Respiratory rates increase in the body's attempt to take in more oxygen. Heart rates increase in the body's attempt to circulate the oxygen around to vital tissues and organs. Blood vessels in the brain dilate. Bone marrow is stimulated to produce a higher count of oxygen-transporting red blood cells. Dense networks of capillaries form, especially at extremities, to increase blood flow, and thus oxygen flow, to the far corners of the body. Oxygen-hungry muscle mass shrinks. All of these physiological responses increase the body's overall efficiency, or ability to do more with less. Amazingly, all of this can occur in a matter of days.

A person noticeably feels only a few of these changes. Increased respiratory rate is most commonly felt in the night, when frequent waking fits of breathlessness punctuate sleep. Increased heart rate often contributes to overall sleeplessness. As more subtle internal changes occur, the symptoms of oxygen deprivation begin to lessen. Headaches eventually dissipate. Shortness of breath fades. Lightheadedness occurs with less frequency. Muscle coordination improves. With enough time, and up to a limited elevation, the human body adapts.

Without proper acclimatization the body goes into decline, and symptoms of acute mountain sickness (AMS) begin to develop. Headaches and dizziness usually come first, often accompanied by lack of appetite or nausea, general fatigue, and sleeplessness. These symptoms are quite common at high altitude, and usually go away with proper attention. Without such attention, things get worse. Loss of muscle coordination, a condition called *ataxia*, indicates a continuing physiological deterioration. At this point the situation becomes serious, and descent is the best treatment. If unattended to, AMS will graduate to either high altitude pulmonary edema (HAPE), or high altitude cerebral edema (HACE).

High altitude pulmonary edema is the most common killer among high-altitude mountaineers. In the case of HAPE, fluid builds up in the lungs and eventually fills them. Early symptoms include gurgling sounds in the chest cavity, chest pain, and extreme shortness of breath. If untreated, HAPE inevitably leads to pulmonary insufficiency and/or cardiac arrest, and death.

High altitude cerebral edema is less common, more difficult to diagnose in its early stages, and is even more deadly. In the case of HACE, fluid builds up between the hard bone of the cranium and the soft tissue of the brain. In its early stages, this often looks like mild AMS, dehydration, exhaustion, or hypothermia, and so HACE is often not recognized until it is too late. The resulting increased pressure on the brain leads to severe headaches, and eventually hallucinations, coma, and death.

There is only one treatment for pulmonary or cerebral edema. It is to go down. Even extreme cases of both diseases may be cured by rapid descent of several thousand feet. Without descent, the diseases are always fatal.

Both kinds of edema occur most commonly at extreme high altitude. But some of the most tragic cases happen much lower, where, unlooked for, they have time to develop into severe cases before they are properly diagnosed. I have seen cases of both pulmonary and cerebral edema at ten thousand feet. Fortunately, in both cases the disease was diagnosed early, evacuations followed, and there were no fatalities.

Mountain people living above ten thousand feet have permanently all of the physiological adaptations that acclimatization provides the lowlander with temporarily, and then some. Their overall metabolic rates can be up to two and a half times those of lowlanders. Their chest cavities are typically enlarged and barrel-shaped to accommodate larger lungs. Their hearts are bigger, beat faster, and last longer. Their red blood cells are more numerous, and contain more hemoglobin. Their capillary networks are denser and more extensive. They are typically of smaller stature, often broad, with a lower ratio of surface area to volume, and are thus more heat efficient. They have decreased sensitivity to cold. They have decreased susceptibility to heart disease, high cholesterol, and most forms of cancer. But there are costs. Mountain people generally have lower fertility and birthrates, and higher incidences of prenatal and infant mortality. They have an increased susceptibility to respiratory diseases such as tuberculosis, and to kidney, gall bladder, and ulceric diseases. They age faster. They generally live shorter lives.

I have heard stories of barefoot Sherpas at eighteen thousand feet, and have seen enough porters in flip-flops on snow with hundred-pound loads to give credibility to such tales. I have seen Sherpa men wake up under the open alpine sky, after having spent the night in nothing but the clothes on their backs, shake off the chill with a smile, and take up their work for the day. I have seen men and women in their twentieth year with as many lines written across their faces as middle-aged men and women of more benign climes. I have felt comfortable in the company of every Sherpa I have met because we have shared the same short, compact physique. The toughness of these people, and their suitability to the high mountain environment in which they live, cannot be singly attributed to attitude or evolution, but rather to a sinuous combination of both.

An hour goes by and still I am on the steep talus and snow. I've crawled faster than this. I still have halfway to go. The pass still eludes me, but I am certain now that we are on route. Several times I look up and see the inspiring sight of windblown prayer flags strung across a small gap in an otherwise impenetrable rock wall. Clouds are billowing and filling the basin below, pooling in the hollows of the landscape below the mountain barrier. I check my watch. It is twenty minutes past eleven. I promise myself if we are not on the pass by one o'clock we will turn around. Although the map indicates relatively mild terrain on the east side of the pass, my faith in its accuracy is now shaken, and I do not wish to tempt fate on the descent. We have underestimated the Renjo La today.

I look at the clouds boiling and pooling against the mountain barrier we are ascending and my heart quails. The familiar feeling of imminent storm is unnerving me. I have a faint foreboding. I feel responsible for us. Then, as if by providence, a small patch of blue sky opens up above the pass. Could the ridge be blocking all the weather and the other side be clear? The thought rings though my head like a bell of hope. I keep moving, with a miraculous new energy. Twenty steps forward. Ten-breath break. Twenty steps forward. A yellow-billed chough casually flies by with nesting material in its bill. I round the corner below the gendarme and a dozen prayer flags come into view, slapping against themselves in the wind and mist. I've never been so glad to see prayer flags.

On the final moves up, I kick steps into eight inches of fluffy new snow with nothing but wet rock beneath. I thrust my leather-gloved hands into the mess, half grab on to the slick, sloped surface of the underlying rocks and pull myself up, up, and up. The prayer flags dance above my brow and nearly touch my head. A whole new world opens up to the east, a world of blue skies and tattered clouds, washed in pure, bright, white mountain light.

The ridge on either side of Renjo La is serrated into lines of angular, broken peaks that together form the formidable mountain wall that the pass so subtly breaches. The ground to the east drops, at first abruptly, then gradually to the tongue of a small cirque glacier. From there it undulates down three thousand feet to the valley of the Dudh Kosi. There, like an immense blue amulet—no, rather a wide-open eye looking up from the face of the Khumbu Himalaya—the sapphire waters of the Third Gokyo Lake stare up at the world in seeming awe. The lake is bounded by the immense lateral moraine of the Ngozumpa glacier, the glacier itself a mile-wide pathway of gray rubble running north-south. Above the debris-obscured ice rises another spur ridge, parallel to this one. Its flanks are shrouded in tattered afternoon clouds and every one of its numerous summits is completely obscured by a thick cloak of gray. Above these growing clouds is a crystalline world of only blue and white. Only one thing penetrates the deep gloom of the cloud banks and rears its brow into the brightness above. It is the striated, pyramidal summit of Chomolungma, Sagarmatha, Mother of the Universe, Mount Everest, glinting in the noonday sun.

River comes up behind me. The last steps up the Renjo La are steep and slippery. As she steps up, I can see that she is crying. But there is joy in her tears.

Renjo La

High mountain terrain

# VIII

The village of Gokyo is situated on the side of the massive lateral moraine of the Ngozumpa Glacier, tucked neatly between the crest of the moraine and the sparkling waters of the lake the moraine is damming. The setting is stunning, especially in the morning, when the new, clean light out of the east lights up the wall of mountains through which the Renjo La finds its way. From the crest of the moraine, above the village, as I look west out over the blue roofs of a dozen lodges, the mountains seem to grow right out of the lake.

I speculate that the village was once (and may still be) seasonal housing for herders pasturing their livestock on the surrounding slopes. It is too high here for permanent settlement, too high even for potatoes. There is no gompa here, no mani stones, no trees, no subsistence agriculture, no trade route, none of the signs of people making a living from the land. But Gokyo bears little resemblance to

Lugare, its seasonal village counterpart on the other side of the Renjo La. There are few traditional stone houses here pasted with dung and mud, roofed with local slate. There are few rock-walled pens and brown-eyed yak calves. Instead, there are large, brightly painted lodges, and their number seems to be growing. There are open-air shops, mountaineering expedition tents in neat rows, and yak trains heavily laden with stuffed nylon duffels and mysterious, large, cylindrical, blue and yellow plastic containers. Gokyo was and still is a seasonal village, but the signs of change are not hard to see. Once it housed seasonal herders. Now it houses seasonal tourists.

Gokyo seemed like Shangri-La from the top of the Renjo La, but today, after nourishment and rest, my relationship with this place becomes more complex. In some ways it is just what I thought it would be: a snapshot in time of a once traditional valley coming increasingly under the influence of Western culture but not quite wholly given over to it. But I was not prepared for just how much Gokyo has already succumbed to Western influences. Numerous sources refer to the valley of the Dudh Kosi as the less-traveled alternative to the Khumbu Valley proper. For countless "by-the-book" trekkers and mountaineers, such words are a seductive invitation, and it seems Gokyo has outgrown itself. It has become a border town in miniature, where traditional culture dwindles in the face of materialism and the inhabitants become hopelessly addicted to the effects of their proximity to material wealth. I am reminded of the dust and filth and confusion and squalor of so many Mexican border towns adjacent to California, Arizona, New Mexico, and Texas. But there are two important differences. First, here, so far away from the noise and convenience of automobiles, the scale is much smaller. Second, I am not sure if the servitude of so many Sherpas is due to a squashed sense of cultural identity or a remarkable transcendence of ego.

The changes brought upon the Khumbu region by westerners has happened over a relatively short time. It has been less than sixty years since Nepal opened its borders to Europeans. When it finally did, the Khumbu region, holding the crown jewel of the planet, was first priority for the Western world. Most reconnaissance expeditions and subsequent mountaineering enterprises focused on the Khumbu Valley proper, and for some decades the adjacent valleys of the region were left alone. Eventually, however, as the Sherpa developed their well-deserved reputation, expeditions to other regions of eastern Nepal sought out their services as guides, sirdars, and porters. Villages along approach paths to the major peaks prospered by opening their doors to westerners, and, like the Swiss of the Alps before them, the Sherpa soon became one of the wealthiest groups of people in Nepal. Eventually the allure of the Himalaya drew an increasingly diverse population of Europeans, and eventually Americans, to the Khumbu region. By the 1960s, mountaineers and surveyors were joined by individual trekkers, organized tour groups, scientists, anthropologists, and artists.

If trends in tourism continue as they have, Gokyo will become another Gorak Shep, the portal village to Everest Base Camp, and Thami will become what Gokyo is today. Only those villages off the main trails (once trade routes, now tourist footpaths) will retain their traditional character. But even those places, such as Khunde, Khumjung, and Phortse, with their commerce tied to Namche, will not go unaffected.

In the late morning I grab my ice ax and head out to the Ngozumpa Glacier for a bit of exploring. As I climb its three-hundred-foot-high lateral moraine, I am struck by its sheer immensity. But this moraine is thousands of years old. The unsorted boulders, rocks, pebbles, and other debris are now covered with a veneer of soil thick enough for rooted alpine plants to grab hold. Grazing terraces line the slopes like a hundred confused trails. The size of this moraine does not reflect the present size of the Ngozumpa, but rather its Pleistocene grandeur. Today, the largest Himalayan glacier south of the main crest is dying in its bed. Its ice is buried in so much accumulated debris, both till and overburden, that if not for telltale escarpments of foliated ice and surficial lakes here and there, one would think this was a gigantic gravel ditch. The ice is so shrunken, so sunken into its bed, that the lateral moraines rise up hundreds of feet on either side. Miraculously, this ice, still a few hundred feet thick in places, will remain here for perhaps hundreds of years, insulated from ablation by so much debris covering its surface, before the valley is emptied of ice and the remaining rubble heap begins to catch soil and seeds.

Descending down onto the glacier's surface is the most treacherous part. The fresh inside slopes of the moraine are at the angle of repose, and rocks and debris trundle downslope at the slightest provocation. After testing out a few spots and getting the same results, I decide speed, rather than care, is the best technique to employ. I run down the debris slope in a descending traverse, dodging dozens of fist-sized chunks of falling rock, and down toward the surface of the glacier. The rocks beneath my feet slide out from under me with every step, but I figure that if I am moving as fast as they are I should be okay. Once I reach the bottom I can still hear rocks moving above. Without looking up, I continue running forward until I am beneath a presently stable section of slope.

The surface of the glacier feels more like walking on a fresh moraine than walking on ice, and for good reason. Till covers the whole thing. There is less than 1 percent of ice exposed. Much of the till is granitic rock, conspicuously different from the metamorphic bedrock of the area. Long ago, when this glacier still moved, the granitic material was transported here, as if by a giant conveyor belt, from further upvalley, where such granitic rocks are common.

The monotony of the undulating hills of till is conspicuously broken by numerous clifflike escarpments of beautifully foliated ice. These outcrops give a cross-section view into the upper layers of ice, laid down hundreds or thousands of years ago like successive layers of sedimentary strata. One outcrop, the largest in the vicinity, attracts my attention. I carefully pick my way across the till piles toward it.

The ice cliff is gray, even its vertical surface veneered by fine debris. Its layers, each a slightly different shade of white and gray, each a slightly different consistency, have degraded differentially, some layers more resistant and forming horizontal ridges, others weaker and forming grooves. In some ways the ice reminds me of the layered slickrock of southern Utah. In other ways it reminds me of banded Lewisian gneiss of the Outer Hebrides of Scotland. But mostly the ice is unique, one of a million expressions of water. Where the layers form shallow overhangs, stalactites of ice have formed, and water slowly drips off the tips of these delicate sculptures. The stalactites remind me of hair, especially the skirtlike hair of the haunches of yaks. The water droplets fall onto other shelves of glacial ice and run down onto still other stalactites, continuing their way down to the partially frozen pools at the base of the escarpment.

Though debris occasionally slides off the top of the escarpment and bounces its way down the face, I decide to take my chances, and I investigate the ice more closely. I walk right up to the ice and run my fingers across it. It is bumpy, not smooth, except where water has run over it and frozen into water ice. The stalactites seem fragile. I run my fingers down the length of one, careful not to break it off. The air is noticeably cooler so close to such a large volume of ice. Curious, I stick my head underneath an overhang, beneath the sharp, dripping ends of the stalactites, and into a shaded recess. It is cold. Icy water drips onto my neck. Sand and small rocks spill down onto my back. I withdraw and move away from the ice.

After some time at the ice escarpment, I climb up onto a prominent hill of till and sit down in the last sun of the day. I can't help but draw parallels between the fate of the ice and the fate of the traditions of the people who live here. Both the Ngozumpa Glacier and traditional Sherpa culture have picked up things along their journey through time that have eventually obscured them. In both cases, the accumulated material protects what it covers, but without rejuvenation from the source eventually what is beneath will shrink and disappear, leaving only its overburden as a telltale sign of its former existence. But in both cases the future is uncertain. Climate change may bring enough accumulation so that the Ngozumpa can flow once again, just as the Sherpa may evolve new traditions, and their changing culture become revitalized.

Raindrops fall on my face, and as I awaken to them I realize I have somehow fallen asleep on a bed of fist-sized granitic till. The debris-covered surface of the glacier, so bright in the late morning sun, has turned gray beneath the ominous clouds of afternoon. Slowly, I rise and begin stepping across the clinking rocks, making my way toward the steep slope of the moraine, back toward the village of Gokyo.

Dying ice

# IX

After a few days of exploring the hinterlands of the uppermost Dudh Kosi valley, we pack up and head east, across the great waste of the Ngozumpa glacier, and up into the high country of the Nyimagawa plateau. We are glad of heart, glad to be walking once again in the mountains, glad to know that tonight we will sleep outside, directly beneath the Himalayan sky.

We depart Gokyo in the late morning, walking south for a mile or so before turning east and crossing the rubble-heaped, silt-covered, surficial lake–studded, dead ice–encrusted, stark, unearthly landscape of the lower Ngozumpa glacier. As we ascend its eastern lateral moraine, the tiny idyllic village of Tagnag (also known as Lagnag or Dagnag, depending on who you talk to), consisting of a few dry-stone buildings and rock walls, comes into view. We descend the moraine and head toward the village for tea.

We enjoy the late morning sun from the patio outside the simple stone structure that is the Cho La View Lodge. The lodge itself is a tiny operation in an equally tiny, quiet, endearing place off the beaten path. The members of the Sherpa family that runs the place are particularly friendly, and we all smile and laugh in each other's company. The mother of the family seems an especially eccentric character, and she makes frequent jokes involving animated charades and expressions. I have never met a Sherpa, man or woman, as outgoing and hilarious as she.

We stay for a long lunch and purchase half a dozen hardboiled eggs and a stack of chipatis as provisions for the next day and a half. As we are preparing to leave, our friend Jeff and his guide, Dawa Sherpa, walk into the village, and soon we are drinking more tea and swapping stories of the last few days' adventures. Jeff's black pants and jacket are conspicuously covered in feathers. He expresses concern about the state of his sleeping bag (rented in Kathmandu), which has so little down in it that it appears translucent when held up to the sun. What few feathers are inside are quickly leaking out through thumb-sized holes in the stitching. He has a plan to rent blankets from the family that runs the lodge otherwise, he speculates, he might freeze. Jeff makes his request to the matriarch of the lodge. She stands next to him, obviously thinking about it, tapping a large ladle against her leg. Then, suddenly, she bursts out laughing and pretends to bonk Jeff on the head with the ladle. The whole family laughs out loud. We all laugh, too. Her answer is no.

We leave Tagnag in midafternoon, giddy from too much tea and excited about sleeping under the stars. We travel along a well-worn footpath, up alongside a rushing mountain stream that is a tributary to the Dudh Kosi. As we ascend, the predictable clouds of afternoon begin brooding over the high peaks and ridges and filling the troughs of the upper valleys. We pass several solitary yaks grazing contentedly on the not-yet-green alpine vegetation. After our ascent of Renjo La and the subsequent days exploring the Dudh Kosi above Gokyo, the terrain seems mellow and the walking easy going. Our bodies feel remarkably different too, finally beginning to acclimatize to the physiological rigors of living above fifteen thousand feet.

After about a mile and a half of ascending, we arrive at the top of a broad saddle. Before us, directly to the east, the prominent north-south trending ridge that separates the valley of the Dudh Kosi from that of the Khumbu proper rises in an intimidating series of serrated peaks breached by narrow cols. It looks just like the ridge of the Renjo La, just like a dozen other spur ridges that jut out southward from the main Himalayan crest. Mist hangs over the broken wall. Snow covers the massive debris cones at its base. Occasionally, the mist parts and we are allowed ephemeral views of the high col of Cho La, the pass we will attempt to cross tomorrow. The pass appears impossibly steep from our perspective. If I did not know it was navigable I would never think to attempt it. I am still getting used to the scale and steepness of Himalayan geography.

We make camp at 16,700 feet, in the shallow alpine valley between the saddle and the ridge of Cho La. There we set up in the dry turf in the lee of some large, mysterious, metamorphic, lichen-covered boulders. Within minutes of arriving, we are snug in our sleeping bags, backs propped up against the boulders, and watching the mist swirl all around us. When the wind is still it is absolutely quiet.

We are visited by a lone alpine accentor, a small robinlike bird, common in the alpine tundra, which has become one of my favorites. Its conspicuous white-dotted wingbars, mouse-gray head and back, rosy sides, white throat patch thinly striped with black, and small, needlelike bill are a telltale combination that distinguish this bird from its redstart and rosy finch neighbors. The accentor hops straight toward us. Watching it closely, we become absolutely still, riveted to our seats. Then, it throws back its little head and emits a gurgling sound from its quivering throat, its bill not even visibly open, and fills the quiet air with song. We are wide-eyed and filled with joy. The accentor hops around camp gurgling and singing for a long while, and eventually takes to singing from atop the large boulder that is part of our camp.

As the evening sets in, we watch, ecstatic, as the light changes the hues of the mountain landscape, mixing and playing with the wafting clouds. I stand and turn ninety degrees with each passing minute, and each time I return to the same view it is completely different in mood and aesthetic. Over the next hour, we are transfixed, and begin a long series of exclamations. "Whoa, look at that!" "Hey, check out that one peak now!" "Ooh! The light just changed again!" "Look how the clouds just enveloped that face!" "Ah! The clouds are purple now!" "Look how crisp that ridgeline is!" Meanwhile, the mist billowing up from the Dudh Kosi condenses a thin layer of dew on our sleeping bags.

Most days the weather does this, but thus far we have spent too many nights indoors, and so have not yet seen the accumulated clouds of afternoon break and scatter in the cooling air of evening. It is perhaps the most fantastic part of the premonsoon diurnal weather pattern, the time in the cycle of clear mornings, cloudy afternoons, and clearing nights when changes in clouds and light happen simultaneously. Sitting at 16,700 feet, amid the cold rock, snow, and ice of the high mountains, beneath the changing Himalayan sky of gray-white, purple, and rose, it occurs to me that the nature of the weather here is inextricably tied to the nature of the mountain terrain, that big, bold, fantastic mountains breed big, bold, fantastic weather.

Without the uplift of the Himalaya, lower Asia would be completely different. There would be no monsoon, and at subtropical latitudes similar to the Sahara, the American, the Australian, the Atacama, and the Kalahari, most of the Indian subcontinent would be a desert, while Tibet to the north would have sufficient water, relatively speaking.

Winters in northern India and southern Nepal would be notably colder without the mountain mass there to block the cold northern air from Siberia. But the point is moot. The Himalaya are inextricably tied to the joining of the Indian subcontinent with Asia. To imagine an Asia without the Himalaya is to imagine Asia without India, and to do so one must reach far back into the deep recesses of time.

The short version of the formation of the Himalaya reads something like this: India collided with Eurasia and the collision pushed up the Himalaya. But a slightly longer version of this unprecedented and fantastic story is much more fascinating.

All of the mountains of the Alpine-Himalayan belt trace their origins back three hundred million years, when the supercontinent Gondwanaland began to break apart. As rifting progressed, eventually India, Australia, Antarctica, Africa, and Madagascar all split off from one another. By eighty-seven million years ago, India was an autonomous piece of crust. It began moving northeast at the breakneck speed of around seven inches a year (incredibly fast by tectonic standards) toward the underbelly of Eurasia. The results were much the same as what was occurring at the same time in the region that was to become the Alps. The once vast Tethys Sea that for millions of years had separated the southern continental landmasses from those of the north began to close up, and its sea floor was devoured.

As India bulldozed its way north, its leading edge became a wedge jammed with the increasingly thick sequence of sediments it was scraping up from the top of the Tethys seafloor. Mixed in with these newly accreted sediments were thin scrapings of volcanic material from the top of the basement of the sea floor. Any islands that punctuated the monotony of the Tethys's abyssal plain were also stacked onto the leading edge of India, sandwiched between the accreting slabs of seafloor sediment. The rest of the seafloor subducted beneath the more buoyant continental crust, and, as was occurring all along what was to become the Alpine-Himalayan belt, broad volcanic arcs formed just inland from the main subduction zone. This same subduction dynamic was occurring along the southern leading edge of

Eurasia, though perhaps to a lesser degree, because Eurasia is not believed to have been moving as fast as India.

Between sixty and fifty million years ago, the vagrant India approached the dominant landmass of Eurasia. Offshore of both of these land-masses, the Tethys had been subducting into a deep ocean trench, and as India neared Eurasia and the Tethys was sandwiched and subducted, the distance between these two trenches became increasingly smaller. When the last sliver of the Tethys went under, the leading edge of India, weighted down by so many accumulated layers of seafloor sediment, seafloor volcanics, and accreted terranes, fell into the Eurasian trench and jammed it. But India did not stop. It ploughed through the immense piles of sediments, accretions, and volcanic arcs, pushing and buckling the sediments into tight folds, shearing them into slabs, thrusting them upward, and smashing the volcanic arcs together. The last waters of the Tethys drained out as its basin was demolished and, over a period of ten million years, from sixty-five to fifty-five million years ago, India and Eurasia were welded together.

The suture that formed as the continental landmasses collided is called the *Indus-Tsangpo Suture*. Today, the line of the suture is described by the Indus River drainage to the west, and the Tsangpo (or Brama-putra) River to the north and east. The suture zone is characterized by a chaotic assemblage of folded, faulted sediments, metamor-phosed, smashed volcanics, and slivers of ultramafic subseafloor rock that were squeezed out during the mêlée.

When India docked with Eurasia, the resulting compression caused the front of India to bow upward into an immense broad arch. As the collision continued, one of two things happened. Some geologists believe that India underthrusted and slid beneath the unmoving mass of Eurasia, doubling the thickness of the Tibetan crust. Other geolo-gists believe that the relatively light rocks of the frontal part of India are too buoyant to have slid beneath the crust of Eurasia, and that the crust of India actually split. The lower, heavier part slid beneath Eurasia, while the upper, buoyant part bowed upward. Recent seismic surveys suggest that Tibetan crust is not underlain by the crust of

India north of the Indus-Tsangpo suture zone, and the front of India has instead formed an immense anticline wedged beneath the front of Eurasia. It is as if India has buried its nose in the dominant crust of the Eurasian continent.

The high Himalaya began to rise by around fifty million years ago as pressure increased between India and Eurasia. With the increased pressure, the material between the Indus-Tsangpo suture zone to the north and the upwarped wedge of the crystalline continental crust of India to the south was squeezed upward. Here were the bent, folded, and broken pieces of the ancient Tethys seafloor. As the squeezing increased, molten material from the mantle intruded into the overlying rocks, forming vast reservoirs of granite and further melting and deforming the surrounding country rock. By around twenty million years ago, the Tethyan sediments and associated intrusions had uplifted significantly and formed a long east-west trending range of high mountains.

In the last twenty million years, more and more of the material between the India-Eurasia suture and the Indian upwarp has been thrust upward, and the Himalaya have continued to grow. In the last three million years, Pleistocene glaciers removed significant amounts of Himalayan rock and redeposited this material in the basins surrounding the range. In response to this removal of material, the Himalaya are currently experiencing the same isostatic rebound that all landmasses affected by large-scale Pleistocene glaciation are. This isostatic uplift, in combination with continued tectonic uplift, indicates that the Himalaya are far from fully developed. When we experience the Himalaya we are experiencing in full swing the largest-scale mountain-building event the Earth has ever known.

As the Himalaya rose, the dominant weather patterns that for millions of years wafted in to Eurasia from southern seas were forever changed. Tibet had long been watered by these southern storms, and its waters drained south, out of the uplands and into the shrinking Tethys. When India docked with Asia, at first the weather systems still penetrated into Tibet, and its southbound rivers rearranged their outlets to find the newly formed Arabian Sea and Bay of Bengal. But as time passed and India continued to grind its way into Eurasia, the Himalaya rose and eventually blocked these storms, casting a pronounced rainshadow on Tibet. The shift happened gradually. Fifty million years ago, when the Himalaya were just beginning to rise, Tibet was not yet a desert. Even twenty million years ago, though much of the southerly weather was intercepted by the growing Himalaya, some of it still made it through to Tibet. But today, with the wall of the high Himalaya reaching well over twenty thousand feet above sea level, precious little southerly weather ever makes it over the main Himalayan crest.

Today, the Himalaya form a physical divide that creates and exaggerates the separation between the subtropical Indian subcontinent and the temperate and subpolar lands to the north. South of the crest, moisture occurs during the summer monsoon, when the Indian plain heats up beneath the sweltering summer sun, and the resulting rising airmasses draw in moisture-laden air delivered by the easterly tradewinds blowing over the Bay of Bengal. As the heat of summer builds, the monsoon grows until eventually the storms pool up against the Himalayan front. The monsoon builds, advancing further upslope as April gives way to May. By the end of May or the beginning of June, only in the earliest morning hours may there be clear skies, and shortly after daybreak the entire Himalaya is socked in. On many days there are no clear skies at all. It rains every day, and the rivers become muddy torrents. Legions of leeches plague the footpaths. The yaks move upslope in hopes of becoming less waterlogged. The mountain people tend their crops and wait it out. Sometime in mid to late September, as the heat of summer relaxes, the clouds break, and the rainwashed skies of autumn ring over the green meadows and snowclad peaks and ridges of the high Himalaya.

To the north, the barren plateau of Tibet knows nothing of these storms. The Himalaya are simply too high for the monsoon to make it up and over. By May the prevailing westerlies have migrated north, and the last storms of winter have passed. The Tibetan plateau basks in bright, relentless sunshine, and scarcely a drop of moisture falls during the time that life needs it most.

In the winter, the weather dynamic reverses, and India dries up while Tibet gets storms. The easterly trade winds move south and cease to deliver moisture-laden air from the Bay of Bengal to India. The Indian subcontinent itself is not subjected to the intense heat of the summer sun, and without the heat pump there are no rising thermal air masses to draw adjacent air in. High pressure settles over India. Meanwhile, to the north, the prevailing westerlies of the temperate latitudes move south, bringing cold air and storms as low pressure settles over Tibet.

In general, the eastern Himalaya experience the monsoon most intensely, and in regions such as Assam, Bhutan, and eastern Nepal, the monsoon is both most predictable and most productive. Here, the influence of winter westerlies is minimal. In the western Himalaya, including western Nepal, Kashmir, and the Hindukush, the monsoon is increasingly less influential, and the winter westerlies are the dominant weather pattern. In the Khumbu, both the monsoon and the winter westerlies are felt.

We have been watching the change from winter to summer, when the effects of the last winter westerly storms mix and intermingle at the onset of the monsoon. Most days we feel the influence of the coming monsoon, indicated by the buildup of clouds throughout the day and predictable afternoon precipitation. On these days the winds blow from downvalley, bringing moist air up from the lowlands. But a few days have been clear all day, and on these days the wind has been out of the west, and the tops of the highest mountains have displayed banners of blowing snow from their glinting summits.

As evening turns to dusk, the clouds that cloaked the mountains dissipate in the purple, blue, and gray light and give way to the clear, cold atmosphere of night. Fortunately for us, the influence of the pre-monsoon weather was not quite enough to bring precipitation to this place, and the surrounding peaks and ridges bear no new snow. This will make for easier going over the Cho La tomorrow morning, and we talk about our route as night falls. One by one, stars and planets begin to materialize above us, and we sit with our heads tilted back, counting and waiting for the next faraway sun to shed its twinkling light on us. It occurs to me then, as the cold night falls and we nestle deeper into our sleeping bags, that in some way I am closer to the rest of the universe than I have ever been.

Porters

Unnamed peaks

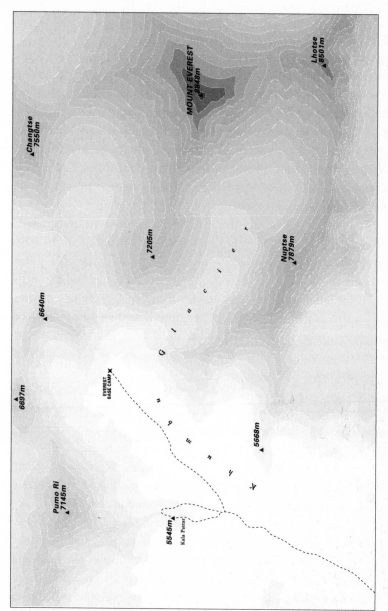

The Everest region

# X

Dzongla is a small seasonal village perched in a hanging valley that is a tributary to the Khumbu. The valley itself is idyllic, a high glacial tread open only to the southeast, surrounded on all other sides by glistening high peaks, cascading streams conjoining into loose braids on the valley floor. The buildings here, however, are broken and encrusted with filth, their insides dark and damp, but somehow still dusty, their windows crammed with soiled, rotting blankets, their inhabitants (Sherpa and westerners alike) unusually somber. The surrounding grounds are littered with piles of trash, toilet paper, and human waste.

The human presence in this valley is a stark contrast to the clean, open feeling of last night's high camp on the other side of the Cho La. Upon arrival we resolved to take tea and meals in the small lodge here but to sleep again in the open. I scouted widely around the small cluster of buildings and found an ideal camping spot about a quarter mile to the southeast. Here, we set up our tent in anticipation of afternoon snow, kicking piles of yak dung aside and singing songs. Then we spread out on the dry, brown vegetation of the meadow to enjoy the last of the day's sun.

Now, snow flurries dance downward from the opaque, pewter gray sky. We lie in the tent, reading and writing and talking. The days seem long with such idle time, and I find myself wishing we spent more time on our feet, traveling across the landscape. But the high elevation still exhausts us, and our bodies cannot live up to the expectations of our spirits, so our walking hours are short. River is afflicted with a chest ailment, which has her congested and considerably weakened.

We are confused about what to do with the remainder of our time here. We fear that downvalley the Khumbu is overrun, and we do not feel called that way, knowing that too much time spent there will leave us feeling empty and dissatisfied. But the valley of Dzongla is a narrow place. We wrestle with our own ethical presumptions about being in the mountains and try to come to terms with what is ahead. River decides that she needs a rest day and will stay in this valley for another night. She makes it clear that she will be fine here without my aid and encourages me to go exploring. After some deliberation I decide to make an unexpected pilgrimage north through the Khumbu Valley to the heart of everything that is contradictory to and difficult to reconcile with my relationship with mountains, to the place more steeped in mountaineering lore, more cluttered by the trappings of modernity, than any other place in the Himalaya: Everest Base Camp.

In the morning I bid River farewell, say my last goodbyes to Jeff and to Dawa Sherpa, and set out, alone for the first time in weeks, for the Khumbu Valley. As soon as I leave the premises of the buildings and begin walking, I become overwhelmed with a feeling of ultimate freedom, and the idea of spending a couple of days solo in the highest mountains on earth becomes increasingly inspiring with every step. I walk briskly along the quiet footpath between Dzongla and Lobuche, traversing high on the shoulder of the Lobuche Ridge and turning conspicuously north and into the Khumbu. In the air, a hundred feet away and right at eye level, a bird of enormous proportions soars high over the valley below. It is too big to be a golden eagle, and too stout to be a lammergeier. Its wingspan must be eight feet, and a full two feet in width. It looks like a giant flying door. Its back is beige, its primaries dark, and its wingtips spread out like black fingers. It is a Himalayan griffon, a behemoth carrion bird, an enormous vulture straight out of a story, a myth, or a song. It peers over in my direction, scans the land, and glides silently off to the north, up the defile of the Khumbu Valley. Never in my life have I seen a bird so big. It is truly Himalayan in scale.

I round the corner and the upper Khumbu comes into view. The valley is broad, obviously heavily glaciated, its bottom filled by the wasting, debris-covered Khumbu glacier, its rims dramatically crowned with an array of serrated peaks. Any remaining traces of color in the landscape are overwhelmed by a stark gray and white of rock and ice. The familiar mass of Nuptse dominates the east side of the uppermost valley and obscures the summit of Everest, which is just northeast of

Nuptse. Northwest of Nupste, the col of the Lho La is the only obvious gap in the wall of the high Himalaya, and the pass neatly frames the pyramidal massif of Changtse, which lies just north of the Himalayan divide, wholly in Tibet. Further west, the white slopes of Pumo Ri glow beneath the only patch of blue in the sky.

My solitude is broken as I join the main footpath of the Khumbu Valley floor and approach the seasonal hamlet of Lobuche. But the traffic is not bad, and I am pleasantly surprised. Lobuche, however, is a hovel. The place seems to spill over with filth. Garbage and dogs surround the dilapidated buildings. Toilets are situated right next to water sources. Everyone seems to be smoking and coughing. The many tales of sickness I have heard from people who stayed here are explained. I don't even slow down, and keep walking upvalley.

The footpath north of Lobuche is good and flat for a couple of miles, then begins a series of ups and downs over the Pleistocene glacial moraines of the Khumbu glacier. Walking over the till is like walking over piles of angular bowling balls. Ahead, the sound of bells betrays a heavily laden yak train. It astonishes me that yaks can make it up and over such terrain, but they do, and their bad steps are less frequent than mine.

At noon I descend to the cluster of lodges that is Gorak Shep. This little hamlet is as far as most trekkers go. It provides access to the relatively low summit of nearby Kala Pattar, which is perfectly situated to offer a stunning view of Everest. Gorak Shep is busy, and dozens of people mill around the premises. I choose a lodge and go in, hungry, and there I purchase extra hard-boiled eggs, spring rolls, and candy bars for an overnight at Everest Base Camp. I rest and sip tea for a long while before departing.

The way to Base Camp is not marked, but it is rutted and easy to follow. I stay behind yak trains for much of the way, listening to clinking bells and the short commands of the herders. Soon the landscape constricts, and heaps of morainal material choke the valley. The pace of the yaks slows but remains steady. I take the opportunity to scamper around them, then tiptoe my way across the till and into the upper Khumbu.

Finally I feel like I am high in the mountains. Here in the Nepal Himalaya one must climb eight to ten thousand feet higher than in the Alps to find a similar glacial landscape, and though the air is noticeably thinner and the mountains on a grander scale, the feeling of being utterly surrounded by the stark, sterile qualities of cold rock and ice is much the same as it is at eight or ten thousand feet in the Alps. From the peaks and ridges of the highest mountains on Earth, a hundred seething blue glaciers emanate, some flowing down into valleys, others hanging on flanks, oozing down in their slow way to form diverse and elaborate lobes and white tongues and bibs, still others broken off at the edges of plateaus, the rock faces below them too precipitous even for hanging glaciers. It is a gray afternoon and the light is flat, which only adds to the drama of the scene by which I am inundated.

From Gorak Shep and the moraines above the hamlet I had been unable to see where the upper Khumbu glacier tumbles out of the Western Cwm and into the valley, and I can understand how from such a vantage point a route up to the upper glacier, and ultimately to the summit of Everest, would have seemed impossible. But I knew such a route was there. Edmund Hillary discovered it in the 1950s and it proved to be the key to attaining the summit of Everest. Now, as I approach Base Camp, I keep looking for the opening between Nuptse and Everest, and even though I know the opening is there, the topography creates such a convincing illusion that it isn't that I begin to doubt myself. Is this really the route to Everest Base Camp? It has to be. I keep walking.

As I near Base Camp I remember that I have no plan, and it occurs to me that though I am prepared, I'm not exactly sure how—or more important, where—I will spend the night. I trust that something unexpected will work out, and if worse comes to worst, I can retreat to some hidden trough in a moraine and bivouac, or walk back to Gorek Shep if need be. Secretly I think that possibly, just maybe, I might see someone I know, or at least someone who knows someone I know.

I turn a corner and am greeted by two men who introduce themselves as the Everest Base Camp welcome party. I instantly recognize
one of them as Lincoln Else, a climbing ranger from Yosemite
National Park whom I have met on occasion in the backcountry of
the Sierra Nevada. He has been invited to work as camp manager for
a small American/Sherpa expedition. In moments, the distance that
separates me from home is reduced from ten thousand miles to ten
feet. Lincoln and I have a dozen mutual friends, have climbed and
explored the same areas of the Sierra Nevada, have read the same
books, and have somehow both ended up here in the Khumbu
Himalaya to meet at Everest Base Camp.

With Lincoln is Andy, a medical student from Oxford, England,
who works in the medical tent at Base Camp. The three of us slip into
easy conversation about the Sierra Nevada, the Himalaya, and other
mountains of the world. They have questions about the state of the
glaciers in the Khumbu and I share with them my rudimentary observations. Within minutes I am invited to Lincoln's expedition tent to
visit and spend the night. Still surprised at the serendipity of our
meeting, I tell them I would be glad to oblige. They give me directions and we make plans to meet there in the evening.

Everest Base Camp is even bigger and more extensive then I expected
or imagined. As I approach the premises, my attention is quickly drawn
away from the innumerable perched boulders and other glacial phenomena to the hundreds of multicolored tents strewn like shanties
across the till-covered surface of the Khumbu glacier. It looks like a
traveling circus. Against the stark whites and grays of the mountain
landscape, the hundreds of tents and thousands of tattered, brightly
colored prayer flags that decorate the camp seem especially vibrant.
There are several huge North Face Himalayan Hotels—geodesic
domes that look like gigantic orange soccer balls cut in half. There are
long tents the size of small houses tied down over stone frames. There
are small personal tents of every conceivable size, shape, and color.
There are specialized shower tents, toilet tents, and dining tents. There
are flags and banners for dozens of expeditions from different countries.

A mixture of excitement and nausea overcomes me. I am overwhelmed. I stop dead in my tracks and just stare, openmouthed.

After some minutes I snap out of it. I locate the white tent with the telltale red cross that is the medical tent, and head for it. As I make my way over the till and ice, I am amazed at the rockiness of the tent sites all around. They all seem occupied, and I begin to wonder where I will pitch my relatively tiny green shelter. After some poking around, I locate a spot that is so good I speculate it must have just recently been abandoned. It is fairly level, and the rocks are only fist-sized. I spill the contents of my pack out on the till and begin setting up camp, just as the afternoon snow begins to fly. With cold fingers, I set my poles together and guide them into their sleeves, then carefully guy the fly out with large, head-sized stones. A lone alpine rosy finch flickers around camp and eventually settles on one of the guy lines, twittering playfully.

As I step back to critically examine the tightness of the tent, I belatedly notice the awesome form of Pumo Ri rising in a nearly symmetrical pyramid into a sky conspicuously devoid of other mountains. Although significantly smaller than its eight-thousand-meter neighbors, Pumo Ri as seen from Base Camp is matchless. From this close up, it is the only mountain you can see in its entirety. And unlike the Everest group of Everest, Lhotse, and Nuptse (the former two of which are not visible from Base Camp), Pumo Ri stands alone. I camp in its shadow.

After some time alone I wander over to Lincoln's expedition camp. It is a modest camp, centered around a rock-walled structure roofed by bright royal blue poly tarps, the same kind you see covering wood piles or old neglected boats in neighborhood yards. These tarps are tied to the rock walls with cheap nylon ropes and the whole rig is held in place by rocks tied or placed along the roof's edge and piled up at ground level to weigh down the overhanging corners of the roof tarps. Surrounding this central structure are assorted tents, the private sleeping quarters of the expedition members. The whole affair has a shanty feel to it that I instantly feel comfortable with. These people are doing more with less and making it work, and I appreciate that.

The camp contrasts sharply with the nearby Ed Viesturs/David Breashears expedition, which is centered around two Himalayan Hotels and neat rows of matching yellow North Face expedition tents. The Viesturs/Breashears camp is the most elaborate, highest profile, and most well-funded expedition at Base Camp this season. Breashears is busy filming for an upcoming movie on the 1996 Everest disaster (chronicled in Krakauer's *Into Thin Air*). Viesturs plans to climb Everest, then get airlifted from Everest Base Camp to Annapurna Base Camp, and from there attempt to climb his fourteenth and final of the eight-thousand-meter peaks.

At Lincoln's camp I am welcomed into the blue tarp-covered central building for tea and conversation. Lincoln introduces me to the small group of people huddled around the table. There with Lincoln and Andy is Linda McMillan, blonde hair, maybe fifty years old, incredibly friendly and welcoming, manager of the expedition, former vice president (and almost president) of the American Alpine Club. Next is Luanne Freer, medical doctor, matriarch of the medical tent, whose dry sense of humor seems well appreciated as a part of the culture here. Last but in no way least is Tom McMillan, brown hair, medium-sized, husband of Linda McMillan, mountaineer and Everest aspirant, who has climbed and explored extensively in North America throughout the last three decades (at least), and once got in trouble with government rangers in the sixties for camping with a bright orange tent. (He can't understand why such high-visual-impact tactics are encouraged nowadays.) Tom and I have several common climbing connections in the Sierra Nevada and Arizona, including both people and places. The most notable of these is Granite Mountain, where I learned how to climb and Tom climbed extensively during the seventies. Tom has a stunning memory, and we end up reminiscing about specific moves on Granite Mountain routes we have both climbed.

Over tea, I get the story behind this small expedition. The story goes back several years, to 1998, when Tom was working for a group called Peak Promotion on an expedition to the South Face of Annapurna.

While on the expedition, Tom met, worked with, and became friends with a promising young Sherpa named Nawang Sherpa.

Two years after the expedition, Nawang was involved in a tragic motorcycle accident in Kathmandu. Nawang lost his leg in the accident, and was sent by his friend and employer Wong Chu Sherpa (of Peak Promotion) to the United States, where Nawang lived with Tom and Linda, received excellent medical care (including fifteen operations and ultimately a prosthetic leg), and attended classes at the University of California. Nawang was fortunate, to say the least. As a capstone project for his rehabilitation, Nawang, working with the McMillans and the legendary American mountaineer and bush pilot Ed Hommer (also an amputee), put together an expedition to Everest titled "Everest: Friendship Beyond Borders Expedition." Before the expedition got off the ground, Ed Hommer died in a rockfall incident on Mount Rainier. For a while the expedition seemed shut down, but Linda and Tom McMillan rallied and put the pieces together. Now, Tom will climb in Ed's stead, alongside Nawang Sherpa, who, if successful, will be the first disabled Sherpa to reach the summit of Everest. They will be supported by Sherpas Nina Tashi and Nina Gombu, who between them have reached the Everest summit sixteen times.

Interestingly, the first disabled person to reach the Everest summit was Tom Whittaker, a British-American who was on the faculty of Prescott College, and a colleague of mine. The connections here seem uncanny.

In the morning the team will begin the first of a series of forays up, through the Khumbu Icefall, to camp one, at about twenty thousand feet. There they will set up a camp and return to Base Camp. After a few days of this they will move permanently, begin sleeping at twenty thousand feet and making forays to twenty-two, and so on, until the final push from camp four, at twenty-six thousand feet. Nawang, who is no doubt ahead of the acclimatization process simply due to his ethnicity, will not be involved with these forays due to his prosthetic leg. He will stay at Base Camp until the others have advanced to camp three, then ascend alpine-style, in one push (sleeping at set camps but continuing upward) to meet the others at the high camp. I quietly wonder how this strategy will go.

Darkness falls, and the full-size Honda generator that provides heat and power to the camp (hauled up by some stalwart yak, no doubt) comes to a sputtering stop for the night. But before we retreat to our sleeping quarters, Tom decides to let us in on a secret. He has devised a sun mask especially for the brutal reflective conditions of the Khumbu Icefall and the Western Cwm. With pride and a healthy dose of good humor, he dramatically removes his newest equipment innovation from his pocket, turns away for a brief moment to put it on, then boyishly turns back toward us. Everyone bursts out laughing. His "mask" is a white T-shirt with eye holes cut out of it, draped over his face and neck. Shouts arise from the onlookers: "Look it's Freddie Kruger!" "No, it's that guy from the Texas movie!" and worse "Hey, they don't allow Klan members on the mountain!" As I laugh, I realize that despite its looks and simplicity of design, Tom's idea is a good one.

The night is cold, and though the silver glow of stars on the white ice of the upper Khumbu is beautiful, I do not linger long, and soon snuggle deep into the down of my sleeping bag. As I lie there with a cold nose, enjoying the feeling of my sleeping bag warming up, I think about this place, these mountains, these people. I reflect on how, in the space of less than a century, Everest and the Khumbu Valley were transformed from terra incognita, the great mysterious unknown, the "third Pole," to the most famous, exploited, and often most desecrated places in the high mountains of the world.

Westerners didn't even know Everest existed until the mid-nineteenth century. It was competition between the northward-expanding British Empire and the eastward-expanding Russian Empire that prompted the British to take an interest in the trans-Himalaya region. At first, Everest was just one of seventy-nine high peaks surveyed by the British from the Indian Plain. It was irreverently called Peak H, then Peak XV. After seven long years of calculating, confirming, and double-checking, in 1856, the surveyor general Andrew Waugh verified that

Peak XV was, in fact, higher than any other peak measured in India, and probably the highest mountain in the world. Eventually, Waugh renamed the mountain after his predecessor, Sir George Everest, who himself never saw the mountain. It was several years before the Royal Geographic Society accepted the name, and in 1865 "Everest" became official. In later years it was realized that the Tibetans already had a name for the mountain, *Chomolungma*, or *Jomo Miyo Lang Sangma*, after the red tiger–riding female deity who lives on the mountain. It was nearly a century later that Nepalis, realizing they had no name of their own for the mountain, gave it its third name, *Sagarmatha*, "Mother of the Universe."

Hidden between the forbidden kingdoms of Nepal and Tibet, Everest remained obscure and unapproachable by westerners for another half century. Eventually, the British, worried about Russian influence in an as-yet-undefiled Tibet, drew back the mysterious curtain that for so long had veiled the central Asian plateau, entered the kingdom of Tibet and, by force of arms, bullied their way to Lhasa. This initially bloody encounter led to the Anglo-Russian agreement of 1907, in which the British and the Russians both agreed to no further incursions into Tibet.

The British invasion of Tibet was led by Francis Younghusband, who years later, in 1921, took the position of president of the Royal Geographic Society. Younghusband, perhaps not yet finished with his conquest of Asia, set his sights on Everest, and soon after assuming the presidency he publicly announced his plans to launch a reconnaissance expedition to the mountain. The man he chose to lead the expedition was the legendary Thomas Mallory.

Mallory was a product of his times. By 1920, the glory days of Victorian England had passed. Dickens, Livingstone, and Darwin had all come and gone. Scott had perished in the Antarctic. The golden age of mountaineering was over. Germany vied with England for ascendancy in Europe. The British Empire began to feel its far-flung lands pulling away from the mother country. Mallory himself was one generation removed from the Victorian era. His heroes were not living men. He had fought as an artillery officer on the western front and so had known the fragility of his culture. He, like Younghusband, saw Everest as a strategic move, a revitalization, a way to set up for a prosperous future, a way to bring back the glory days and show that it was Englishmen alone who had the grit and character to stand on top of the world.

For the three and a half years spanning 1921–1924 Everest drew the attention of the world, and inextricably tied to the identity of Everest was that of Thomas Mallory. Mallory led three expeditions to the mountain, all from Tibet via Darjeeling, all up the Rongbuk Glacier to the North Col, and all unsuccessful. On the first expedition (1921), Mallory's party made twenty-two thousand feet before they retreated due to weather. On the second expedition (1922), they made twenty-seven thousand feet before retreating, brutalized by the cold. A month later, Mallory made another attempt, but met with disaster just below the North Col, where nine Sherpas were swept away by an avalanche, seven of whom met their deaths in a deep, cold, blue crevasse. On the third and final expedition (1924), after an initial attempt, storms, and a retreat to Base Camp, Mallory was blessed with a break in the weather. They made the North Col, and, climbing with young Sandy Irvine this time, Mallory made Camp V, then Camp VI. On the morning of June 8, the two men walked across the summit ridge, were enveloped in a cloud of mist, and perished on the mountain.

It was nearly three decades before the British actualized their vision of being the first to stand on top of the world. In the interim, Tibet closed its borders to foreigners. Nepal, meanwhile, was opened to the west. Exploratory expeditions led by H. W. Tillman, among others, paved the way for subsequent expeditions, which focused on the peaks themselves. Paramount among these was the British Everest expedition of 1953, led by John Hunt. Hunt, however, is seldom remembered for the expedition he so expertly led. Instead, we remember the expedition for the two climbers who were the first to undisputedly make the summit of Everest. Both men climbed without incident and lived to tell the tale. Those men were Edmund Hillary and Tenzing Norgay.

Hillary's and Norgay's names have become synonymous with that of Everest. Hillary instantly became the most famous mountaineer ever known. A New Zealander, Hillary climbed for the British Crown, and upon his success he was quickly elevated to the status of international hero. He made the front page of every major newspaper in the Western world. He was knighted. In his home country, he became a cultural icon. His face gazes out from the front of the Kiwi's five-dollar bill. Norgay found less fame abroad but more at home. To this day, the image of him atop Everest can be found on the walls of almost every lodge throughout the Khumbu region. Countless young Sherpas hold Norgay up as the ultimate inspiration. His progeny find cultural recognition and livelihood in his legacy. Both Hillary and Norgay lived up to their reputations as local heroes by dedicating years of service in establishing schools, infrastructure, and livelihood for native inhabitants of the Khumbu region.

When I think of the history of Everest, three images come to mind. The first is Mallory, together with Irvine, two black specks on a distant ridge being enveloped by clouds. The second is young Hillary and Norgay standing on the summit together beneath the crystalline blue sky of the stratosphere. The third is a lone man, unaided by Sherpa guides, climbing partners, ropes, or supplementary oxygen, who has faced the forbidding, massive nature of the Himalayan giant of giants solo and returned unscathed. That man is Reinhold Messner.

Messner has become synonymous with Everest because, like the mountain itself, he is a singular figure. Unlike many before him, Messner dedicated his *entire* life to the pursuit of alpinism, and thus has became the archetype of pure focus. He began climbing in the Tyrolean Alps at the age of five. By the age of twenty he had climbed over five hundred routes in the Alps. By age twenty-five he had climbed all of the classic north faces in the Alps, including the Matterhorn, the Eiger, and the Grandes Jorasses, many of these in winter and/or solo. His first

Himalayan climb was the Rupal Face of Nanga Parbat, the highest big wall in the world. In 1978, after successful climbs of Mansalu and Hidden Peak, and a second, solo climb of Nanga Parbat, Messner, along with Peter Habeler, became the first to climb Everest without oxygen. The next year he reached the summit of K2 without oxygen. Then he returned to Everest and climbed solo, without oxygen, from Tibet up the Rongbuk Glacier to the North Col, and on to the summit.

What set Messner apart from his contemporaries and will forever make his accomplishments above and beyond the scope of his successors was his style. He stood in stark contrast to the knighted, iconic figures of his predecessors. He was not wealthy or particularly well educated, and he was slightly awkward looking, with long hair and a big, shaggy beard. Messner challenged the dogma of the Himalayan mountaineering scene by employing fast, lightweight, alpine tactics to the mountains of Asia. His expeditions were small and under-funded. He never used fixed lines, and rather than relying on the skills and hard work of Sherpa guides, he did everything himself. He abandoned the use of oxygen as unnecessary and unethical. Unencumbered by large parties and elaborate equipment, he was able to move quickly, from Base Camp to summit, with only hasty bivouacs in between. He was bold, to be sure, but his success rate, as well as the fact that he is still alive today, indicate that Messner was more than just lucky. He possessed an uncanny ability to make good decisions and to survive in the face of extremely adverse conditions.

Messner's accomplishments only began with his pioneering solo ascent of Everest. In subsequent years he went on to become the first person to climb all fourteen of the world's eight-thousand-meter peaks. He became the third person to climb the highest mountain on each of the seven continents. He crossed Greenland the hard way (south to north), by foot and sledge, solo. He traversed the Takla Makan Desert. He was the first person to cross Antarctica on foot, with no technical support. Somehow, in between these trips, Messner has written over forty books on his adventures, the proceeds of which provide funding for his expeditions.

Mallory attempted what no one else had. Hillary and Norgay accomplished what Mallory couldn't. Messner accomplished what Hillary and Norgay never dared to try. Each of these characters pushed the envelope further than the last, expanding our collective human consciousness of what is possible. After Messner, one wonders what is left to do, if it is time to hang it up, leave the giants alone for awhile, try something different, maybe just stop and take a look around.

Thousands have lost their lives on the flanks of the big Himalayan peaks. Piles of discarded oxygen tanks, tattered remains of tents, countless pieces of discarded equipment, tons of human waste, and long-frozen bodies litter the vicinity of Everest and the other thirteen eight-thousand-meter peaks, as well as dozens of lesser giants. Scores if not hundreds of people flock to the summit of Everest each year, many of them wealthy enthusiasts who pay their way onto expeditions and are guided, and in some cases nearly dragged, to the top. Although the 1996 tragedy on Everest brought the dangers of such commercial ventures to the attention of the world, nothing has changed. The gains of expanding the scope of human potential become obscured in avalanches of commercialization and self-absorption.

The costs are incalculable. Many are critical of the apparent irreverence for the sacred nature of mountains that seems inherent in the activity of mountaineering. Krakauer summarized these sentiments in his careful use of a quote from a young Sherpa whose family died in the service of Himalayan mountaineering expeditions:

> *I am a Sherpa orphan. My father was killed in the Khumbu Icefall while load-ferrying for an expedition in the late sixties. My mother died just below Pheriche when her heart gave out under the weight of the load she was carrying for another expedition in 1970. Three of my siblings died from various causes, my sister and I were sent to foster homes in Europe and the U.S.*
>
> *I never have gone back to my homeland because I feel it is cursed. My ancestors arrived in the Solo-Khumbu region fleeing from persecution in the lowlands. There they found sanctuary in the shadow of "Sagarmathaji" [Sagarmatha], "mother goddess of the earth." In return they were expected to protect the goddesses' sanctuary from outsiders.*

*But my people went the other way. They helped outsiders find their way into the sanctuary and violate every limb of her body by standing on top of her, crowing in victory, and dirtying and polluting her bosom. Some of them have had to sacrifice themselves, others escaped through the skin of their teeth, or offered other lives in lieu.*

The native inhabitants of the Himalaya never climbed mountains. It is a practice of the West, where the essence of life that is found only in the face of danger has been eliminated by technology, medicine, and other insurances, and so we seek out danger as a kind of sport. Somehow, perhaps unknowingly at first, our attempts to experience the lost essence of life has had lasting effects on the landscapes that have become our pleasure fields, and the cultures that inhabit them.

Western mountaineering, and subsequently trekking, has so completely altered the economy of traditional Sherpa culture that along main approach routes one must look hard to find an adult whose livelihood is not connected to the regular influx of westerners to the area. Family structures are reorganized to account for men being gone during the pre- and postmonsoon seasons. Sherpa men compete fiercely for the top mountaineering positions, and prestige at home is largely based on their accomplishments in the field. Yaks no longer bear trade goods from Tibet and the Indian subcontinent, but rather gasoline-powered generators, plastic drums full of human excrement, computers, radios, and the assorted junk of the West, little of which ever goes back home with those who brought it, but rather collects in heaps and covers the ground. Stone and mud houses of ancient design are remodeled into brightly painted lodges complete with modern Western facilities such as running water, flush toilets, showers, and even hot tubs. Gone are the age-old traditions of transhumance, polyandry, and strong local identity. These have been replaced by service industries, Western cultural norms, and increasing homogeneity. In the face of all this, somehow the opportunistic nature of the Sherpa, which is itself a long-standing tradition, allows them to adapt, as they always have, to changing times. But this change, like the culture that brings it, is bigger and faster, and I can only hope these people's cultural identity will endure.

Most westerners I have met in the Himalaya believe that, because westerners have brought money into the relatively small economies of the local inhabitants of the Himalaya, change is a good thing. This is the same rationale that is often used by people who hire porters to carry their loads for them. But there are some assumptions in such a rationale that are important to recognize and challenge. First, that money is good. We in the West have shown that it is not. We have demonstrated that it divides people, does not unite them, that it stratifies otherwise egalitarian societies unnecessarily, that the quest for its eternal increase leads to cancerous economies dependent on growth at the expense of resources that are ultimately finite. Second, that bigger is better. But smaller economies are proven without a doubt to be more sustainable, happier, healthier economic atmospheres for people to live in. Third, that keeping one culture in servitude to another is somehow good for the culture that is in servitude, as long as they are paid a fair wage. In short, these rationales fail to take into consideration the integrity and long-term well-being of the local people, and give westerners a long list of excuses to keep doing what we are doing and somehow feel good about it.

I linger on the edge of sleep, unable to attain its bliss in the thin, cold air, unable to detach from the waves of frustration that are the product of my thoughts. The Khumbu glacier booms and pops and groans beneath me, sending formidable echoes throughout Everest Base Camp. Sometime in the night, a huge icefall releases from the south face of Pumo Ri and sends an avalanche of ice trundling downslope. It sounds like thunder but keeps rolling and rolling for well over a minute. In my half-sleep I panic, and for a moment it seems as if the avalanche will consume Base Camp and bury all of us alive. But the booming passes, and eventually I drift into a fitful, dreamless slumber.

Everest Base Camp and the Khumbu Icefall

# XI

The instant the sun's rays strike Everest Base Camp, the temperature soars scores of degrees in a span of a few minutes, and whatever accommodations one has made for the bitter cold of night quickly become a suffocating, sweltering overburden. Everyone awakens simultaneously and an unintentional ceremony occurs. People emerge from their tents with tea and sit out in the brilliant sun, staring aghast at the illuminated spectacle of the high Himalaya that all but engulfs this place. Even people who have been here for weeks still sit awestruck in the morning. The High Himalaya casts a powerful spell.

I rise and look to the east, to the jumbled frozen cascade of glacial ice that is the Khumbu Icefall. Without breaking camp, I grab my sweater, my sunglasses, and my ice ax and start making for it. I meander through a dozen expedition camps and pass a hundred people enjoying the morning ritual of sitting and tea drinking and admiring this supreme mountain landscape. Everyone is in a good mood in the morning here.

In the light of the day, the blue glacial ice encrusting the surrounding slopes gleams brilliantly, and the white of the snow surface that obscures so much of the underlying ice is intense and hard to look at. As I pass the last brightly colored nylon tents, I leave the till-covered part of the glacier on which Base Camp sits and become surrounded by glacial ice. There are standing waves of ice as high as a person. There are strange melt patterns in the surface that resemble desert badlands. There are mounds of bare ice that rise up out of snow-covered flats. The forms here resemble nothing I have ever seen in temperate mountain regions, and I can only attribute the obscurity of the melt patterns on the surface of the upper Khumbu Glacier to the high-angle intensity of the subtropical sun. Of all the times of the year to be here, now, during the premonsoon season, is the prime time to view these forms. In the summer they will be covered with snow regularly, and through the fall the accumulated snows of the summer monsoon season will slowly but surely melt out. In the winter they will intermittently be covered by winter storms and swept clean by wind.

I follow the obvious route through the icefall, and as I ascend, the ice becomes noticeably more fractured. The unfamiliar melt forms of the flats are replaced by more familiar seracs, broken chunks of glacial ice ranging in size from refrigerators to cars, typically rectangular and standing up on end. The ice is so broken here because the bedrock beneath it is steep and irregular, and the plasticity of the glacier near its surface is not enough for the ice to bend around the bedrock. In some way the glacier is like a Slinky going down a staircase. In order to go down, its outer side has to expand and open up. Like the Slinky, less expansion is felt on the inside of the glacier, and cracks rarely penetrate into the ice more than a hundred and fifty feet. Here, the Khumbu glacier tumbles out of the Western Cwm above and drops around two thousand vertical feet in one mile. The ice can't bend that much, so it breaks into thousands of pieces, all of which somehow tip and trundle downslope, riding on the more plastic ice beneath, and pack back together below for the relatively smooth ride out of the Khumbu Valley.

The Khumbu Icefall is a difficult section of the Everest climb because the broken nature of the ice, in combination with the steep angle of the slope the glacier descends, means that seracs of all sizes can unpredictably break and fall at any time, smashing whatever is immediately below them before they themselves are smashed and ruined. Nowadays, each season the expeditions at Base Camp cooperate and agree on a route up the icefall, complete with fixed rope lines and aluminum ladders where needed. The route, of course, changes every season as the ice moves and changes. This strategy alleviates some of the danger of moving through the icefall, but certainly not all. Many experienced mountaineers anxiously tiptoe through this section of the climb, even after a few passages through on acclimatization forays.

I continue upward, taking time to investigate particularly shimmering outcrops of blue ice, peer into freshly riven crevasses, and finger the metamorphosed snow crystals on the surface that will soon contribute their mass to the greater glacier. I pass several fixed lines on my way up, and at first I use them more out of habit than necessity, but soon that changes. As I explore, I become aware of several climbers descending the icefall above me. They look hot and tired, dripping with sweat, encased in full mountaineering regalia from head to toe. They look at me strangely as they pass by and offer nervous greetings. Seldom do anyone but climbers venture into the icefall. Somehow I'm sure it is illegal for me to be there. Suddenly I realize it is most likely my appearance that beguiles them. I am in sneakers, pants pulled up to my knees, shirtsleeves, and I have a wool sweater wrapped around my head like a turban. I laugh out loud as the climbers clamber down the ice.

Eventually, the angle increases such that progress without crampons becomes genuinely difficult. Around five hundred feet above Base Camp, I turn for a look down at the spread of the glacier, then up at the mass of Pumo Ri that opposes the Everest group. The scene is spare, simple but not without subtlety, elegant. It seems to clear the mind and sharpen it to a keen edge, like freshly hewn stone of a mountain or crystalline ice, like the diamond cutter the Buddha spoke of, capable of cutting through the seemingly impenetrable constructs of attachment and desire. I take in a breath of thin air, offer my acknowledgments to this supreme landscape, and head back down the ice.

In Dugla, River is awaiting my arrival, as expected, at the first teahouse. Here we share tea and stories for an hour and make plans for our descent from the Khumbu. We decide that rather than stay in Dugla and walk by day, we will leave immediately, walk during the evening hours, and so avoid traffic on the footpaths.

It is late afternoon by the time we set out and resume our walk. Once we drop down from Dugla, the valley opens to an expansive flat-bottomed glacial trough with an impressively high Pleistocene moraine on its eastern side. We cross the floor of the valley, and despite the incessant wind we can still hear the roar and gurgle of the braided, pale, cobble-strewn meltwater stream of the dying Khumbu glacier.

We pass through the busy village of Periche but do not stop. South of the village, we cross the river and follow the footpath that works its way up the western bank, avoiding a section of narrow gorge where the valley constricts and the river downcuts vigorously. We traverse high on the west bank for a mile or so as dusk descends on us, and the summit of Ama Dablam comes into view mysteriously, abruptly, awesomely to our east. From afar, we see the welcoming green roof of a lone lodge and we know it will be our place of rest.

In the morning we continue our descent. While walking, I realize that even in the short timespan of weeks I have gotten used to the high mountains. Once below fifteen thousand feet, the air itself seems noticeably more luxuriant, almost liquid. We walk southward and into the valley breeze, through the hillside village of Pangboche, down to the river, across it, and up toward the Tangboche Monastery, the largest and most prominent of the temples of the Tibetans-in-exile that exists in the Khumbu region. In my mind I can't help but think that this place, like Everest Base Camp, is a place apart, a place that, like Base Camp, reflects the worldview of the people who consider it holy.

The footpath takes us through a luxuriant mountain forest of stunted juniper, wine-colored birch, fragrant fir, and brightly blooming rhododendron. I am struck by the richness and beauty of life; the contrast between the cold, gleaming towers of ice and the sweet, swaying trees that surround me seems so very stark. All of the signs of permanent human settlement have returned: green, soft trees, carefully carved prayer stones, fertile fields, smiling women, red-cheeked children. All of this after a few weeks in the sterile world of the high mountains is enough to make one whistle, sing, tuck away for a quiet nap in the copse of an overhanging juniper, and be glad to be alive.

A final steep hill brought us to the frequently visited monastery of Tangboche and the cluster of lodges and shops that have predictably sprung up around it. The monastery instantly strikes me as beautiful and complex, not unlike the Khumbu Glacier. While the beauty and complexity of the glacier is reflective of elemental nature, so the beauty and complexity of the monastery is reflective of human nature.

The Tangboche monastery is as ornate as any Catholic church I have seen, and it is full of the idolatry I have come to associate with Tibetan Buddhism. The buildings are the typical squares, with brightly colored window frames, and overhanging roofs on all sides. The front gate is guarded by two larger-than-life-size stone lions and an elaborately carved stone gateway. The main gompa is surrounded by what appear to be living quarters for the monks in residence. Several monks pull potatoes from the small plots between the buildings.

I take my shoes off and go inside the main gompa. Although I have seen similar places before, I am nevertheless instantly struck by the level of intricacy and detail of the interior of the room. A single monk sits at a low table in the center of the room, fingering prayer beads and muttering an inaudible mantra. Elaborate silk banners form into vertical cylinders of all colors dangling from the ceiling, decorating the space of the room. Long, low tables covered with maroon robes and drapes and pillows provide sitting room for a hundred monks. The walls are painted meticulously with thousands of figures and symbols and stories, all of which seem to depict the pervasive

Buddhist theme of human suffering. There is a man having the flesh carved off his body and put on a scale to be weighed against his bare skeleton and head. There is a man helping someone across a river, only to have his eyes gouged out by the man he aided. There are hundreds of images like these, and also life-size depictions of Sakyamuni Buddha and other important Tibetan figures whose origin remains mysterious to me. But these images pale in comparison to the one on the west side of the room. Here, surrounded by lesser figures, is a twenty-foot-high golden statue of the Buddha, replete with a studded topknot, heavy-lidded eyes, and a double chin.

The level of idolatry is shocking to me. On one hand it is a fascinating display of the blending of Buddhism with the older, animistic B'on traditions of pre-Buddha Tibetan culture. On the other hand it is a grotesque departure from the central teachings of the Buddha, riddled with attachment to ritual colors, forms, and hierarchy. I leave the gompa confused. Why do religions always seem to immediately depart from and contradict the teachings that are most central to them? There was Jesus Christ and there was the Catholic Church. There were the Ten Commandments, then the institutions of the Pharisees and Sadducees. There were the Vedas, then the complex hierarchies of Hinduism. There was what the Buddha taught and there is what the Buddhists do. It occurs to me that the West often looks with envy at the religions of the East, as if the traditions of the East are somehow more pure or offer insights that the West missed out on. But today it occurs to me that Christ and the Buddha sit eye to eye, and the traditions that bear their names drift equally far from their sources. The twenty-foot statue of the Buddha here at Tengboche is no different from a thousand twenty-foot-high crucifixes in a thousand churches. Instead of a Buddhist monk sitting and reciting mantras, there would have been a Christian priest uttering Hail Marys while fingering a rosary. Instead of brilliantly colored paintings on the walls, there would have been stained glass, meticulously depicting the stations of the cross.

While it has been said that the East values enlightenment and the West does not, I see also a place where everything is sacred so nothing is, where castes still live on in the names of people, and where the sheltered monk is considered closer to God than the pious man who pulls potatoes from his dry fields in the flying snow, cooks soup for his family and guests, and every day of every week of every month of every year still offers the first food or drink of each day to the god who lives right there, in his own house, and the one just outside his window, and the gods everywhere.

Khumbu ice sulpture

I leave Tengboche confused and shaken. We descend through the steep forest to the river, then climb back up again and head for the markets of Namche.

Tengboche monastary

# XII

Namche. The sounds of men working stone resounds through the air like a hundred broken bells clinking without rest from sunrise to sunset. Shopkeepers haggle with travelers over yak bells and beads and Buddhas and fists full of rupees. Chickens scuttle beneath overturned packbaskets while zopkioks and dzums command the right-of-way in the narrow lanes. Pictures of poly-bags of rice and lentils and greens and old water buffalo carcasses mix with images of plastic sandals and cigarette smoke. The hum of local commerce, full of vitality, shows no sign of wavering, reaching across town to where the trail drops into the forest of scented blue pines, and winds its way to the river, to the green hills, to the jungles of the Terai, to the Indian Plain, or up to where the footpaths lead to the highlands, up and over hulking glaciers, to the plateau of Tibet. Namche is the meeting place to which all paths lead, the trading grounds, the bottleneck, the boutique, the bazaar, the barter town. As such, it is here that we return before bringing our trip full circle.

Only after so much time in the high mountains am I ready to appreciate this crazy mix of a place. I wander through the shops and talk to the local merchants as best I can. In the afternoon, I sit and drink cup after cup of hot, sweet milk tea. Over the course of a few hours, nearly every person I've met over the past few weeks passes through the lanes of town. When I greet them, one by one, we invariably smile and laugh out loud at our reunion.

In the morning, River and I head down the dusty footpath, through the pine forest, and down to the crashing gorge of the Dudh Kosi. We weave our way through the upcoming foot traffic. Tired porters strain against tumplines, their feet cracked and splayed, sandals thin and torn, followed by clean and confused-looking westerners equipped with hydration systems, and hundred-dollar trekking poles clinking against the stones. Brown men with empty baskets are our company on the way down, the empty baskets indicative of a good day at Namche's Saturday morning bazaar. They walk with the purposeful but relaxed gait of those who are returning home after a long time away. We walk with a similar gait, but slower, as do those who may be seeing things for the last time. We absorb every rhododendron blossom along the way, and watch leaf warblers glean insects from the trees, and sunbirds draw nectar from flowers. Before reaching the valley bottom, we find a secret trail down through the thickets to the cold river. There, we swim in the cold, milky-blue waters of the Dudh Kosi amid a thousand multicolored dragonflies.

Monjo: rain on spring green fields of buckwheat, the sounds of small children playing in narrow, wet, stone lanes, the smell brown, moist, aromatic soil and pine wood burning in stoves, the sights of white wild strawberry blossoms on the woodland floor. Monks play horns and drums in a small outbuilding as the rain falls softly on metal roofs. I sit in a wooden room, sip tea, and prepare to move my pen across a blank page. I remember sitting here just weeks ago, pouring over maps and trying to envision the experiences of the weeks to come. Now we have lived the map, and its lines and colors have become rivers, mountains, glaciers, fields, and villages filled with living people. As I reflect, I am strangely brought to my little wooden house in a faraway land across the wide sea, where I have sat countless times over a steaming cup of tea and scratched new lines across wide-open pages. There is no place like this place, but some things are just the same. Outside the open window, the sounds of the rain mix with those of the small children playing in the street. I remember a passage from the Tao Te Ching that reads, simply, "To go far is to return home."

Namche Bazaar

# THE LAND OF THE LONG WHITE CLOUD

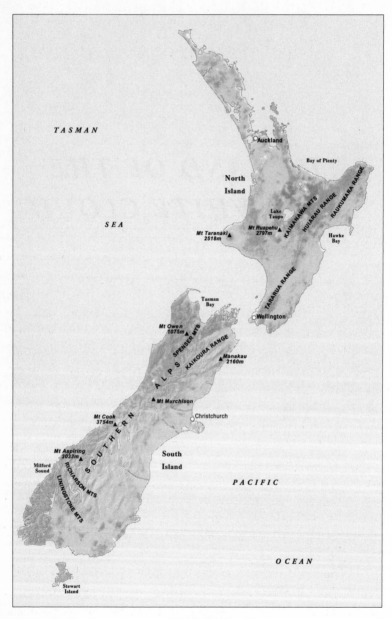

New Zealand

# *The Southern Alps of New Zealand*

## I

**THE SOUTHERN HEMISPHERE** is an ocean planet. It is nearly 90 percent seawater. It is possible to circumnavigate the world in the south and never even see land. The vast majority of the Earth's landmasses are clustered around temperate latitudes of the north, including all of Europe, all of the Asian mainland, all of North America, Greenland, the bulk of Africa, and part of South America. Africa and South America straddle the equator and both make significant appearances south (though in the case of Africa only one-third of the continent is southern). Uniquely southern are Australia, Indonesia, New Guinea, New Zealand (often collectively and conveniently referred to as *Oceana*), scattered islands of Polynesia, and ice-mantled Antarctica. Separated by so much ocean, those landmasses of the south that do not straddle the equator are some of the most geographically isolated landmasses in the world. They have not been connected to the lands of the north for hundreds of millions of years. Since then, the evolution of species in the lands of the south has gone on unhindered, unchallenged, and unimagined by the living things of the north. Going there, to those far-flung islands of the Southern Ocean, is about as close to going to another planet (without leaving the planet) as one can get.

The Southern Alps of New Zealand are unique among the mountain landscapes of the south because they are the only alpine mountains that occur on such a geographically isolated island. Both New Guinea and South America host higher mountains, but in the case of New Guinea, these are tropical mountains, lacking expanses of alpine tundra and the effects of glaciation, and in the case of the Andes of South America, the continent is vast, straddles the equator, and though it was once geographically autonomous, it is now connected to North America. To see the southern analog of the great ranges of the north, and to see such mountains in true geographically isolated glory, one must go to the storm-beaten, temperate, rainforest-cloaked, metamorphic-dominated, glacier-encrusted Southern Alps. And to understand the Southern Alps, one must understand the land in which they are set: far away New Zealand, the last inhabited place on Earth.

Yesterday I flew over the Pacific Ocean for twelve solid hours, and during that entire time there was no land in sight. After a brief stop in Fiji, a large island by Polynesian standards but still a tiny pimple in the expanse of the Pacific, the plane ascended once again and flew west, away from the pink glow of the tropical dawn. Now, thirty thousand feet below, the blue of the Pacific dominates the sphere of the Earth as far as the eye can see in all directions. The monotony of this surreal scene is broken only by occasional puffy cumulus clouds thousands of feet below, which cast shadows onto the surface of the ocean. I scan my map of New Zealand before me and look out the window anxiously. This clear day is a blessing. With so much ocean around it, New Zealand is often mantled in clouds for days on end. But not today. Soon the living map will appear below, the best geographic introduction to New Zealand I could hope for.

The islands of New Zealand trend north-northeast to south-southwest and are over three times as long as they are wide. They spread their length across nearly fourteen degrees of latitude, reaching from a subtropical thirty-four degrees in the north to a subpolar forty-eight degrees in the south. As such, they are nearly eight hundred nautical miles long, around two hundred miles wide, larger than all of the United Kingdom, and a little smaller than California. The South Island of New Zealand alone is about the size of Nepal and four times the size of Switzerland. These are enormous islands by Pacific standards, dwarfing any other landmasses between Australia–New Guinea and the distant Americas by ten or a hundred times.

The next time I look out the window, a gigantic swath of green land bound by tan beaches cuts a conspicuous slice into the field of blue ocean. The land below is way too big to be any of the tiny islands between Fiji and New Zealand. I glance at my map and realize that it is the North Cape of the North Island, reaching out like a finger groping curiously from the main mass of the island. We fly over the west coast of the cape, heading due south. Soon the green expanse of the entire North Island is below, replete with neatly cultivated fields, wide expanses of pasture, thick woods, and conspicuous, massive volcanoes breaking the pastoral monotony. The North Island is primarily a volcanic landscape, formed at the convergent boundary of the Australian Plate and the Pacific Plate. It is a classic island arc of volcanoes, like Japan or the Philippines. The eroded material from the flanks of successive volcanoes has come to rest around the still active cones, forming a broad, undulating topography with luxuriant beds of fertile soil.

The west coast of the North Island ends at the Cook Strait, which separates the two islands of New Zealand. We cross over the water and to the indented coastline of the Marlborough Sounds region, the north coast of the South Island. As we continue south over the vast green of the Nelson country, I can see mountains in the distance, and they are big mountains. Their peaks are snow-covered, and they extend in parallel rows to the south-southwest. They go on for miles before they become engulfed in white, glowing clouds. My eyes widen and I find myself pressed up against the window. The South Island is a decidedly mountain landscape. The Southern Alps define the island physically, biologically, and culturally. The entire length of the island is essentially one long mountain range, or series of parallel ranges. The main crest of the range rears up from the west coast in a dramatic tectonic display of rugged topography. To the east the country is more gentle, a broad plain composed of sediments washed down from the mountains over the last few million years.

As the plane continues south, it, too, becomes engulfed in white clouds.

Surrounded by so much ocean, it is no wonder New Zealand was the last major landmass on the planet to be discovered and colonized by human beings. Long after Australia, the Americas, the remote regions of the Arctic, and even the far-flung islands of Polynesia had been discovered, New Zealand remained unknown to the peoples of the world. Eventually it was the Polynesians, those expert and profoundly bold seafarers of the Pacific, who came across New Zealand. Polynesian legend tells of a skilled navigator named Kupe, who reached New Zealand from French Polynesia in 950 CE Kupe explored the coasts of New Zealand and the interior of the North Island before returning to his people. The name he gave to the Islands was *Aotearoa*, the *Land of the Long White Cloud*.

It was centuries before Polynesians migrated to Aotearoa. The journey was long and risky, and as time wore on, Kupe's discovery of the vast lands south and west faded into legend. But legends do not die; rather, they grow in the imaginations of people. By 1350 CE, when the Polynesians became gripped by population increases and the corresponding depletion of resources, they recalled the legend of Aotearoa, and as they had done countless times before, they took to the sea. A small fleet of great canoes put their bows into the Pacific foam and set a course that followed Kupe's centuries-old instructions. No one knows exactly how many boats set out, but Maori tales tell of ten boats that landed in New Zealand. To this day, all Maori trace their origins back to one of those ten boats.

At the time of the Maoris' arrival, the islands of Aotearoa had never known mammals except for two species of bats. Left unmolested by the mammalian predators so common on the mainland continents, many of the native birds of Aotearoa grew to enormous proportions and lost their ability to fly. The vegetation, having been isolated from mainland competition for hundreds of millions of years, had evolved in slow motion, and ancient tree ferns long extinct stood sixty feet tall in the wet forests. The arrival of the Maori heralded the first of a series of changes that would transform the composition and structure of the Aotearoan ecosystem forever. They brought domesticated dogs and pigs with them. Rats made it over on the boats. They brought yams and other domestic crops that they planted widely in the more tropical Polynesian islands. They hunted moa and other enormous flightless birds even to extinction, following the last of these large, easy prey into the far reaches of the islands and thereby spreading out across all of Aotearoa.

Due to geographic isolation, Maori culture evolved for hundreds of years uninfluenced by outside cultures. Though the temperate climate of New Zealand was new to them, they made up for what they could not grow by augmenting their ages-old agricultural practices with hunting. They remained a Stone Age people, due to a combination of lack of competition with other cultures and lack of tool-quality metals. They did, however, practice agriculture, had elaborate social systems, and waged both ritualized and real warfare extensively. Each of the ten canoe groups established their own exclusive territory, and tribal battles and wars over territory followed.

Three hundred years after the Maoris' arrival in Aotearoa, the first Europeans discovered the land they came to call New Zealand. The sixteenth, seventeenth, and eighteenth centuries were a time of tremendous power struggles among the European nations. England, France, Spain, and Portugal all vied for dominance over trade routes to Asia and the Americas. The spice trade, the slave trade, the fur trade, and the first efforts toward colonization of new lands were in full swing and regulated the economy of the western world. The Dutch made it into the milieu, not through strength of arms or wealth in lands, but rather though their business savvy. They formed the Dutch East India Company, and by the mid-seventeenth century had become a commercial force to be reckoned with.

The Dutch explorer Abel Tasman left Jakarta, Indonesia in 1642 in search of new lands and potential trade routes in the south. After circumnavigating the Australian continent, he sailed east and across the wide sea that now bears his name. Eventually, and unexpectedly, Abel Tasman discovered the west coast of Aotearoa. However excited about his new discovery the Dutch navigator may have been, the native

Maori were less so. When Tasman landed, the Maori promptly killed and ate several of his crew. Tasman was shaken by this gruesome encounter, and he promptly withdrew, sailing up the coast and back to Jakarta. The Dutch never returned to Aotearoa, but they gave it a name of their own, New Zealand, after the Dutch province of Zeeland. It was not long before the entire Western world learned of its existence.

Miraculously, no Europeans went near New Zealand for over a hundred years after Tasman. It was the British explorer Captain James Cook who finally rediscovered the Islands. From the time of the ancient Greeks, Western knowledge of geography had given rise to speculation that the world must be balanced, and that land must be distributed roughly equally across the globe. But as the explorations of the sixteenth and seventeenth centuries probed further and further south of the equator, they revealed that there was little land "down under." European scholars, unable to yield to the amassing empirical evidence, postulated that there must be a massive southern continent south of Australia. Tasman's discovery of New Zealand was a tiny piece of such a continent, and since Tasman had never sailed around the islands, there was no reason to think anything different.

Cook dispelled the myth once and for all. In 1769, he circumnavigated New Zealand, thoroughly exploring the coasts along the way. He mapped both of the main islands, naming numerous features along the way. Finally, unbeknownst to the native Maori inhabitants, Cook claimed the whole place for the British crown. He eventually sailed south, in 1774, penetrated the Antarctic Circle to beyond seventy degrees latitude, and finding only water there, disproved indisputably the existence of a large southern continent (Antarctica, still yet to be discovered at this time, is much smaller than geographers once speculated.) In 1777, upon his return to Britain, Cook published the accounts of his voyages. In that year, the teeming millions of Europe learned of the green lands of New Zealand, and the gaze of Britain turned south.

The coming of Europeans, like the coming of the Maori before them, brought big changes to New Zealand. Within fifty years of Cook's publication, the Maori culture was nearly destroyed. Sealers and whalers quickly decimated the marine mammal populations around the islands, and introduced smallpox, measles, mumps, influenza, syphilis, gonorrhea, prostitution, metal, guns, and Western greed to the native Maori. The subsequent arrival of Christian missionaries resulted in more subtle devastation of the Maori by eroding traditional Maori spiritual beliefs, practices, and lifeways. Although the missionaries' efforts helped curtail the devastating effects of disease, the Maori population continued to decline. An increasing stream of European settlers bound for already occupied territory only exacerbated the situation. Violence erupted as Europeans and Maori vied for land. The British, in an effort to stay potential French colonization efforts in New Zealand, kept encouraging British colonization. New Zealand was in an uproar.

In 1840, the British officials in New Zealand drafted a treaty that ensured unqualified exercise of chieftainship of the Maori over their lands, people, and possessions, as well as equal rights and citizenship with all citizens of England, provided the Maori relinquish sovereignty of the land to the Queen of England. After some minor objections and amendments, the treaty was eventually signed by over five hundred Maori chiefs throughout New Zealand. The treaty became known as the Treaty of Waitangi, and its endorsement by the crown on February 6, 1840, marks, for some, the birthday of New Zealand.

What seemed initially like a good deal to the Maori soon became a nightmare. Like so many well-intended treaties between indigenous peoples and newcomers, without the ability to enforce the treaty the Maori quickly crumpled beneath the increasing weight of British colonization. Full-scale war broke out in 1860 and raged for over five years, and the Maori, badly outnumbered and lacking modern weapons, were almost entirely wiped out. The British government confiscated almost all of their remaining lands, and in 1877, exactly one hundred years after Cook's publication, the British declared the Treaty of Waitangi null and void. Today, over three million people of European descent live in New Zealand, while the Maori number less than five hundred thousand. Waitangi Day, February 6, is celebrated widely by those of European decent, but for those of Maori blood it is a bitter day.

In the seven hundred fifty years since humans first occupied the geographically isolated New Zealand, the land has seen many changes. The great moa, eight feet tall and several hundred pounds, are but ghosts in the green forests. The mice and rats of the Maori have wreaked havoc on countless species of small, ground-nesting, flightless birds, who themselves evolved over eons and eons. The Maori, in turn, came near to extinction when, for the first time after centuries of cultural isolation, they were forced to compete with another culture. Much of the original forested landscape has been replaced by field and farm, town and city. Conservationists go to great pains in attempts to eradicate introduced mammals and plant species native to the Northern Hemisphere. But New Zealand is still an incredible land, unique, novel, breathtakingly beautiful, its people friendly beyond expectation, and having made some amends, peacefully coexisting with one another. And today, over a millennium after Kupe's initial discovery of Aotearoa, it still lives up to its evocative name, the Land of the Long White Cloud.

The world has gone gray and the plane is suspended in an amorphous shroud of clouds. As we pass through this thick veil, raindrops strike my window and leave long streaks of glistening water across the otherwise clear glass. Below, the city of Christchurch slowly comes into

view, at first in elusive glimpses, then as an expansive whole. It is blurred by the water streaks on my window, but I can tell nevertheless that it is big. Too big. My first priority is to find a fast way out of town and into the mountain landscape of the Southern Alps. According to my map, the place where the mountains come closest to Christchurch is called Arthur's Pass.

Kea

# II

The Southern Alps above Arthur's Pass stand cloaked in heavy clouds bearing moisture from the surrounding oceans. Their burden runs down deep gorges of crumbling graywacke laced with bright green ferns. The water cascades at relentlessly steep angles, threading around tree roots, free-falling in a frothy whitish-gray color that matches the sheen of the clouds. Everywhere I go, the sound is inescapable—the sound of the sea trying to get back to itself. How many times does a water molecule get captured by the air, drawn up into the sky, blown over land, condensed into a cloud, rained or snowed down upon the earth for a long journey through glaciers and lakes and churning streams, always back to its source, the sea?

The mountains stand in defiance of the forces of water and gravity, which together work to reduce all landforms to flat plains level with the sea. Here in the Southern Alps, the crust of the earth is squeezed up and twisted, bent, and folded into mountains of youthful relief and vigor, still in the throes of being created. Here, somehow, the opposing forces of uplift and degradation are both found in extreme forms. Steep, rugged mountains. Relentless rain and snow. Today, it seems as if the weather works harder, rains more, and somehow the mountains rise even faster, higher, and will not be taken down. I sense a tenacity in these misty mountains. Though they are rugged they are not made of hard stone. They are mostly soft, composed of crumbly graywacke and unstable, only lightly metamorphosed schists. This is one of the few places in the world where a person can sit and watch geologic processes happen in the space of a few hours; when the water washes over the bones of the mountains you can literally watch the place get worked down. Streams cut deeper gorges daily. Braided outwash rivers push gravel into heaping bars, and the river is rerouted into new channels. Small and even large landslides are regular occurrences. Yet somehow, despite all this, the mountains persist in their growing.

It is late morning now and I walk purposefully through the thick woods on the north flank of Avalanche Peak. As I pass through the tree tunnels and head up the steep, slippery slope, I watch the water from the ocean and from the sky, tear the place down. As I watch the forest floor zoom by beneath my feet, I notice that the tree roots, like threads, hold the ground together and keep it from washing away entirely. But the water also forms threads, and these pull the slope apart. The threads of water seem to be gaining. The track I am walking on has been eroded to a three-foot-deep trench cut into the steeply angled forest floor, revealing polished graywacke bedrock where the soil has entirely washed away. Unfamiliar birdsongs drift through the air and mix with the sounds of water. Mysterious flowers poke out of hollows on the forest floor. Strange trees form a thick canopy above me.

The trees end abruptly at just over four thousand feet above sea level. It is about the same level as treeline in the Swiss Alps, or the northern Appalachians of New England, or the Coast Ranges of British Columbia, all places of exceptionally rugged subpolar climate. I find myself in a vast and open landscape thickly carpeted by unfamiliar grasslike species, unmistakably monocots. As I ascend, the cloud level lifts ever so slightly, revealing the wide, gray valley bottom of the Waimakariri River to the south, and a continuation of forest-clad mountains to the north.

I ascend the open, steep ridge rapidly and soon overtake the slowly lifting clouds. As I enter into them, the world again becomes veiled. Periodic breaks in the swirling mist reveal the beautiful curve of a ridge extending south to the next peak south of Avalanche, Mount Bealy. I hope the clouds clear so I can make the traverse. The tussock vegetation soon gives way to more familiar prostrate alpine vegetation—mats and cushions—and immediately I feel as if I could be in the alpine zone anywhere in the world. But the band of such vegetation is narrow and gives way to the bare, crumbly, dark gray rock of the summit ridge. I scramble across the ridge and make for the silhouette of a prominent gendarme. Once I am past it, the ridge ends and drops noticeably away on all sides. If not for the ground close at hand I would have no idea I was even on the summit. I am completely engulfed in cloud.

There are other subpolar mountain ranges that stand in close proximity to oceans, face the prevailing weather systems, and so have wet maritime climates. The many ranges of the western North American Cordillera, ranging from the Sierra Nevada of California to the volcanoes of the Aleutian Islands, are among these maritime mountains, as are the mountains of Norway, the Scottish Highlands, and the Patagonian Andes. But of these many ranges, only two, the Patagonian Andes and the Southern Alps, have oceans on *both* sides. Of these two, only one is literally surrounded by ocean. Today, for example, the weather does not come from the west, off the Tasman Sea, where it almost always comes from. Today it is unusually clear on the west coast on the South Island of New Zealand. Today the weather comes from the east, off the South Pacific. In just about any other maritime subpolar mountain range, if the weather were clear in the direction of the prevailing systems, the mountains would be standing high and dry. But not here. Instead, the Southern Alps are drenched in rain. In the hamlet of Arthur's Pass, three thousand feet lower than the summit of Avalanche Peak, clouds cover the sky for more than 250 days of the year. Rain falls on most of those days. The high mountains surrounding the pass are socked in even more often.

The extreme maritime position of the Southern Alps not only brings abundant moisture to the region, it moderates the climate considerably. Temperature extremes are relatively low. At sea level, for example, it rarely snows, and only does so much farther south. Temperatures on the west coast, which is most influenced by maritime conditions due to its position facing the prevailing westerlies, rarely dip below freezing in winter or rise above eighty degrees in summer. Only in the mountains does it get truly cold, and here the snow falls in abundance. Typical of maritime regions, the snow is usually heavy, with a high water content. In the southern winter, it piles and heaps over the Southern Alps.

Because the majority of the weather at subpolar latitudes comes with prevailing westerlies, the Southern Alps, like all of the subpolar maritime ranges of the Earth, capture much of the moisture in the westerly airmasses, and cast a rainshadow over the lands on their lee side. To the east of the Southern Alps, the Canterbury Plain basks in a relatively high number of clear days. Although the continental influence is minimal in a place as maritime as New Zealand, the lands to the east of the mountains do experience noticeably greater temperature extremes, both on a diurnal and seasonal basis. For perspective, the Patagonian Andes cast a similar rain shadow over adjacent Argentina, and while the west coast in covered in thick temperate rainforest, the Argentinian plain is covered in semiarid pampas grassland. In North America, the western Cordillera, well watered along its western front, creates the Great Basin Desert and vast, dry interior conifer forests to the east. Even Scandinavia receives significantly less snowfall in the interior of the mountains.

When the prevailing westerlies reach the mountain front, these moisture-laden airmasses are eventually forced to rise up and over the mountains. As an airmass rises, it cools. Air can hold increasingly less moisture in vapor form as it cools, and as the temperature drops, moisture vapor condenses out of the airmass and becomes liquid water droplets. As these droplets converge, they form clouds. As the droplets continue to converge, they coalesce and grow heavier. Eventually, when they become heavy enough for gravity to work on them, they fall out of the cloud, and it rains or snows.

A rising airmass will always cool at a rate of five degrees for every thousand feet. But as moisture vapor condenses and water changes state from vapor to liquid, latent heat is emitted. Condensation actually warms the airmass, offsetting the normal cooling rate by about two degrees for every thousand feet. Moisture-laden airmasses, then, because they are always condensing as they cool, cool at a lesser rate of three degrees for every thousand feet, while dry airmasses will cool at the greater rate of five degrees per thousand feet. In spring or fall, for example, a moisture-laden airmass off the Tasman Sea may be fifty degrees when it reaches the west coast. As it rises over the eight-thousand-foot crest of the Southern Alps, it cools and condenses its moisture out, sending a deluge of rain down on the lower slopes and heavy snowfall in the higher mountains. Cooling at a rate of three degrees for every thousand feet, by the time the airmass reaches the crest of the mountains it is twenty-six degrees. Cold. As the airmass *descends*, however, it warms, and no longer condenses its moisture out. Thus, it warms at a rate of five degrees for every thousand feet. In addition, as it warms, the airmass has an increased capacity to hold moisture in vapor form, so it draws moisture from the surrounding landscape. This warming, drying wind is sixty degrees when it reaches the Canterbury Plain. While the west coast is socked in by maritime conditions, the lands in the lee of the mountains enjoy interior continental conditions.

But it is rarely that simple. Today, the lands to the east look like the hills of Ireland on a bad day, while the west coast skies look like those of a southern California beach town.

The coldest weather, as one might expect, comes from the south. Southern winds blown in from Antarctica can cause temperatures to plummet in the mountains. As in other regions of the world, these relatively cold temperatures are usually associated with relatively clear weather. Cold air, even when it moves over a thousand-mile expanse of seawater, can only hold and transport so much moisture.

As I sit amid the swirling clouds, I am visited by an enormous green parrot. It is the size of a raven, and seems equally curious. Its black bill gives it away. The upper mandible is huge and sharply hooked, extending in a steep decurve from its nostrils, and is at least twice as long as its lower mandible. It hops over to me for an inspection and is soon joined by a second bird. They are *keas*, both native and endemic to New Zealand, and from what I have read they are at least as intelligent as ravens, and may be the most intelligent birds in the world. I begin speaking to them out loud, and I can tell from their actions that they are intrigued and seem to know that I am addressing them directly. I explain to them that I have no offerings of food for them today, that I will not be accepting any from them either, and that it is for the best for all of us. They seem undeterred. Our interaction continues for half an hour until finally I stand up to depart and they flicker off into the pumping clouds.

Wilberforce River

# III

The Waimakariri River is so big that the thought of trying to ford its many braids seems like suicide. The regular route up the river valley stays along the riverbed for about eight miles to the main fork in the upper river, necessitating anywhere from a dozen to a hundred river crossings. Under the present conditions, even a single crossing is out of the question. The water itself has turned from familiar blue-green to a nearly opaque chalky gray. Mixed in with the sound of rushing water are the sounds of head-sized boulders saltating along the riverbed. Rivers kill more people every year in New Zealand than any other natural phenomena, rivers such as the Waimakariri, on days just like this one.

My plan is to traverse the main divide of the Southern Alps. It is a route that will take me along the river valley of the upper Waimakariri, over a series of high mountain passes, and eventually out the west-flowing Styx River, named for the Greek's legendary river of the Underworld. Along the way I will pass through beech forests, ford swollen streams, scramble and slide through rocky outcrops, negotiate snowfields, traverse expanses of alpine tundra, and end up in the thick of a temperate rainforest. I will witness firsthand the transition from the east slope of the Southern Alps to the west slope. If the weather pattern stays the same, I will gaze on the briny waters of the Tasman Sea as they lap unfamiliar shores.

My map shows a track that follows the south bank of the river and I decide to try it. As I leave the road and enter the luxuriant forest, I am filled with fresh excitement. Big drops of water fall from the canopy above me. The sounds of moving water surround me. Too soon, I emerge from the forest and traverse the wide delta of the Jordan River, a southerly tributary to the Waimakariri. Its two-mile-wide delta seems disproportionately large for the size of the actual stream, even today, when all waters are in flood stage. As I walk across the plain, the clouds lift, revealing the massive form of Mount Stewart to the northwest. I spook several hares out of the brush and they hop and dart ahead of me, then disappear under cover. They were brought here, likely by the British, for sport hunting. There are no four-legged mammals native to New Zealand.

After crossing the Anti Crow River, another southerly tributary to the Waimakariri, the track gets worse. It enters the thick beech forest and seems to appear and disappear at random. In many places the Waimakariri has undercut the banks (which the track follows) and slick, broken graywacke crumbles away to open air, with the raging river twenty feet below. I cling to tree trunks and tiptoe along, sending rocks and soil to disappear in the swift river. The vegetation becomes so thick and the trees so close I have to put my head down and force myself and my pack through, meanwhile negotiating hidden waterfalls and small but treacherous landslides. As I plow through the vegetation, the leaves release their burden of water droplets down my back and I am flooded by rainwater. I have to use all four limbs to negotiate such terrain.

The forest here at nearly three thousand feet above sea level is a montane forest, but it is starkly different from the montane forests of the Northern Hemisphere. Pine needles do not cover the forest floor. There is no familiar scent of spruce or fir here. In fact, none of the twisted trees even have needles. Instead, they have tiny, round leaves. In my mind, the montane forest clear up to treeline has always been defined by the presence of conifers, and in some places, birches. I realize that unconsciously I have come to equate the forested parts of mountain landscapes with the smells, textures, sounds, and wildlife that are associated with the conifers of the family *Pinaceae*. The thought never even occurred to me that there could be mountains without pines, firs, spruces, Douglas firs, or hemlocks. But there are no conifers like those of the north here. These trees are of a completely different lineage. They are southern beech trees of the genus *Nothofagus*, and I have never seen the likes of them in my life. Deer, bear, and marten do not lurk here. Instead, strange hardwoods with tiny leaves grow thick. The place smells sweeter and wetter than I am accustomed to. Birds can be heard, but the quiet presence of large animals is altogether missing. It is not that they are gone, but rather that they have never been here.

A hundred million years ago, New Zealand was still partially connected to Gondwanaland, that great southern supercontinent that included South America, Africa, Australia, India, and Antarctica. At this time, much of the southern lands were cloaked in forests of southern beeches of the genus *Nothofagus* and primitive southern gymnosperms of the genus *Podocarpus*. Vast tropical lands or, in some cases, oceans separated these forests from those of the north, in which the family *Pinaceae*, with its genera of *Pinus* (pines), *Abies* (firs), *Picea* (spruces), and *Tsuga* (hemlocks), developed and diversified. Mammals had not yet evolved, and flightless birds roamed the primeval forests while ancestral penguins inhabited the coasts.

Around eighty million years ago, Gondwanaland broke up, and South America, Africa, Australia, India, Antarctica, and tiny New Zealand all began divergent tectonic trajectories. As each of these landmasses became increasingly geographically isolated from the others, the species they had once all had in common began to evolve and diversify according to the new and changing environmental conditions of the lands they inhabited. New Zealand rifted relatively early, still before mammals evolved, and as it rafted out onto the wide seas, the possibility of subsequently evolved mammals ever making it to such a faraway place dwindled to nearly nothing. Australia rifted shortly after, but in the interim, primitive mammals had developed, including monotremes and marsupials, and when Australia left Gondwanaland, it was with representatives of these mammalian orders on board, which in time evolved and diversified to fill the continent. Both South America and Africa eventually came into close enough proximity to the northerly landmasses that more recently evolved mammals moved in. Of Antarctica, we may never know what is under the ice.

Today, New Zealand, southern Africa, South America, and Australia (especially Tasmania) all host beech trees of the genus *Nothofagus*. In each of these diverged landmasses, unique species within *Nothfagus* have developed. In South America, for example, the beeches are primarily deciduous, while everywhere else in the south they are broad-leaved evergreens. In New Zealand, four distinct species of beech are recognized, each with characteristics best suited to a specific habitat type. In general, all of the southern beeches occur in temperate regions of the Southern Hemisphere and prefer moist habitats. In New Zealand, beeches grow both in mixed forests and, especially at montane elevations between three thousand feet and treeline, form pure stands.

Here, along the Waimakariri River, the beeches grow especially close together, unlike in the more open beech forests typical east of the great divide. I can only attribute their relative density to the amount of rainfall here, most of which seems to be funneled directly down my back. I am amazed by how much water I am experiencing.

After two miles of some of the worst forest travel I have ever experienced, I emerge from the bush and drop down to the open gravel flats of the riverbed. Relieved, I follow the valley bottom north, crossing over the many braids of Greenlaw Creek, and make good time for the next two miles to the confluence of the White and the Waimakariri rivers. Here, I look on the west bank for the welcome shelter of Carrington Hut. It is there, tucked in the trees, and though I have mixed feelings about things such as huts in the wilderness, today the hut is a blessed sight. I move in for the night, and within an hour I am cuddling a pot of hot soup.

I awake to gray and the promise of another day of rain. It rained so hard through the night I thought the hut and everyone in it might float away down the flooding Waimakariri all the way to the Pacific Ocean. Today I am supposed to go above the bushline, up and over two high passes and above the permanent snowline. Over breakfast, I entertain the idea of heading back out the river valley and satisfying myself with a few days of day hiking around Arthur's Pass. But the rain stops, and I conjure some hope. I resolve to resume my transalpine route, to leave immediately, and to try to be over both Harmon and Whitehorn passes before the rain starts again. In fifteen minutes I am outside, walking.

I follow the bank of the White River upstream to the spot where the Clough Cableway crosses the river. The cableway consists of a beat-up old open aluminum box-frame car big enough for one or two people to stand in, suspended by three drooping steel cables. The car is moved over the cables to the other side by the passenger(s) turning a hand crank that powers a set of geared wheels. The cables span the three-hundred-foot-wide river gorge some hundred feet up off the ground. I look down at the river and decide that today the cableway is the only way across. I step into the car and start cranking. It is hard work, but one look a hundred feet down at the river crashing over boulders is enough to keep me going. I land safely on the other side and send the car back across the river.

I ascend the narrow, rocky gorge of the Taipoiti River, which is really a high-gradient mountain cascade. As I move up the gorge, steep graywacke walls laced with ephemeral bridal veil falls loom up on either side of the down-cut streambed, and tufts of bright green mosses and clusters of tiny white composites peek out from small cracks and ledges. After over a thousand feet of climbing and several stream crossings, the sides of the gorge open up and the slope's gradient lessens. As I make the final steps to the top of Harmon Pass, the familiar feeling of open alpine country refreshes me. Still no rain.

As I rise up over the broad pass, I am afforded a brief view up the narrow glacial valley that leads to Whitehorn Pass. The pass itself is difficult to make out, as the white of the permanent snow blends almost imperceptibly with the clouds. On either side of the pass, however, the twin peaks of Isobel and Rosamond are unmistakable, and these two stand as waymarkers for my intended route over the divide. Clouds swarm around the whole scene, threatening to engulf the pass and the peaks around it. A cold, gusty wind blows downvalley. I shelter behind a boulder just long enough for a quick snack and some water, then I shoulder my pack and head up and into the white.

Soon, I am hopping along large, pale gray boulders, coarse glacial till from ages hence. The movement is very familiar to me, and for the first time since my arrival in the Southern Alps, I begin to feel good and truly comfortable in my environment. Boulder hopping eventually brings me to the snow, and just as I step out onto the gently sloping white field, the clouds swarm in around me. The whole world goes white, and I lose all sense of depth perception. I just keep putting one foot in front of the other and climbing. Although visibility is poor, traveling is easy and my ice ax remains tied to my pack. The snow angle is low and the days of rain have created a soft surface layer of snow to kick steps into. Slowly, methodically, I plod my way up.

Eventually, the snowfield ends, and I am confronted with what seems in the mist to be a long, precipitous drop off the opposite side. I have come up in the wrong spot. To my left the snow meets dark gray rock that rises abruptly into the wind-torn clouds. To my right the snow follows a low-angle gradient downslope and disappears into the swirling mist. I sally back down a bit and follow the edge of the snow to my right. It leads to an obvious low point that I decide must be the pass. There, I find a boulder to crouch behind and eat, drink, and put on more clothes. The wind is bone-chilling. Tiny raindrops fleck my face. I do not linger, and head forward, off the snow, onto uncertain rock, and into the mist.

As I descend, the rain comes down. Everything about the landscape seems exceptionally dramatic because of the ever-changing visibility. Black walls and steeply angled heaps of rock appear one minute and are gone the next. Snowfields materialize from out of the clouds and are enveloped in seconds. My map indicates the presence of the Cronin glacier, just feet away, hanging in a prominent cirque on the south flank of Mount Rosamond, but I never see any sign of it. Eventually, immense scree fields and cryptic peaks give way to the green vegetation and waterfalls that are the headwaters of the Cronin River. I keep to the left of the stream, at times hopping stream cobbles, at other times waist-deep in tussock and other mysterious vegetation. As I near the confluence of the quickly building Cronin and the larger Wilberforce River, the Cronin cuts into a steep, nonnavigable gorge. Here, I cross the Cronin and follow a steep route up and onto the right-hand bank.

The rain stops suddenly and I watch, transfixed, as a vague hint of blue develops and opens up over the upper Wilberforce Valley. It stops me in my tracks. The sunlight seems to pour through like a warm, golden liquid. Across the valley of the Wilberforce, the snow-clad southern face of Mount Beals glints in the light. Below it, the white, pulsing cascade of Hamer Falls drops three hundred feet to the valley below. Even as I stand there, the mountains seem to cast off their burden of clouds and, for the first time since my arrival here, stand revealed in light.

I catch my breath and walk on, descending to the confluence of the Cronin and the Wilberforce. There, on a small grassy flat just above the river bar, is Park Morpeth Hut. It is a small eight-by-twelve-foot box sided with green-painted corrugated steel. The bright red door gives the place a festive feel, and is welcoming. Inside, the hut is dark and damp. There are four wood-framed bunks with brown burlap stretched over them, and a small table at the lone window. As the clouds once again envelop the landscape and raindrops spatter against the windowpane, I boil up a kettle of water and settle in for tea. I give thanks for the small window, and I stare out at the world longingly.

Out on the small grassy flat beside the hut, a brown bird the size of a lean chicken emerges from the bush and starts poking around the building. It has an odd walk, like that of a rail, and appears off balance, its feet placed too far back, its body leaning forward like a rounded cone tapering to an all-purpose, raven-size bill. But its most astonishing feature is that it is obviously flightless. Its wings lack flight feathers entirely and are reduced to seemingly useless appendages, not large enough or developed enough to hold aloft a bird half its size. It seems curious and unafraid, as if it knows this place well and is a common visitor. I frantically thumb through the pages of my bird book, trying to do so without taking my eyes off of the bird.

New Zealand has been likened to an ark of Gondwanaland. As such, when Gondwanaland broke up, New Zealand carried off "specimens" of whatever grew, walked, swam, or flew on Gondwanaland at the time. Flightless birds made it on board, mammals didn't. While nearly everywhere else in the world flightless birds were wiped out by subsequently evolved mammalian predators, in New Zealand, because of its geographic isolation, these birds, along with hosts of plant species, were preserved over countless millennia. Eventually, in such isolation, many of the species of the "Gondwanaland Ark" began to speciate and diversify, but without the competitive pressure of subsequently evolved species of plants and animals, many of the primitive traits first developed in Gondwanaland persisted, and still do today. In short, New Zealand is a kind of biological museum of Gondwanaland; it is as close as we may ever get to experiencing the Earth as it was eighty to a hundred million years ago.

But evolution did not stop in New Zealand, it just took off in its own direction. In the case of the North Island, not much of it even existed when Gondwanaland broke up, and as repeated episodes of volcanic activity built it up, pioneering plants and animals enjoyed a colonization free-for-all. In such rare cases when new habitat literally bubbles up out of the brine, or, as more commonly occurs, when existing habitat is devastated by glaciation, volcanic activity, or any kind of disturbance, the slate is clean for new biota to move in. Ecologists refer to such situations as *ecological releases*, and the term refers to *any* time in nature that competition for niches is removed, be it by new habitat being created, disturbance cleaning the slate, or mass extinction wiping out entire taxa of organisms in a single blow. Quite simply, the supply of ecological niches exceeds the demand, rather than the other way around, as is the case in stable ecosystems. As long as supply exceeds demand, competition for niches is nil, and the question becomes who can get there first, rather than who is the best competitor for a particular niche. With no competition from specialized competitors, opportunistic, novel, and just plain "weird" species, many of which have been overshadowed by more dominant species prior to whatever change precipitated the ecological release, have a chance at making it, and evolution gets liberal for a time. Once all the niches are filled, specialization begins to occur, and evolution settles back into a conservative pace. When the North Island of New Zealand began stacking up, evolutionary liberalism had a field day.

For perspective, most of the big leaps in evolution that we are familiar with happened because of ecological releases. A mass extinction of marine invertebrates and fishes during the Devonian Period resulted in the radiation of insects and amphibians. A mass extinction of all species of animals at the close of the Permian Period resulted in the radiation of reptiles and the earliest mammals. The mass extinction of the Cretaceous Period led to the extinction of the dinosaurs and resulted in the large-scale radiation of mammals. In each case, the slate is swept clean and new varieties of life are afforded a chance for dominance that they would never have been given if they were not released, for a time, from competition with specialists. In the case of

New Zealand, a similar release from competition took place when the island rifted from the mainland of Gondwanaland, and occurred again as new lands piled up above sea level.

Isolation leads to preservation of primitive traits and release from competition with increasingly specialized competitors. The existence of new, previously uninhabited lands leads to further ecological release and the opportunity for still further novelty to develop. The result? Nearly all birds native to New Zealand (except seabirds) are endemic, meaning they occur here and nowhere else in the world. It is a birder's paradise, where almost every bird seen is a completely new experience in life. Among plants, of the two thousand species of flowering plants (this does not include numerous species of tree ferns and podocarps, among others) over 75 percent are endemic. It means the plants and animals of New Zealand are especially reflective of the place in which they live. They define the place. They embody its essence.

But it also means that New Zealand's native species, for so long geographically isolated, are especially vulnerable in a world of fast-paced ecological collisions. Not just the species, but the ecosystems themselves are particularly fragile. The great moas that once roamed the islands were hunted to extinction by the Maori in the space of a few generations. The Maori themselves nearly suffered a similar fate. Today, the more specialized species of flightless birds, such as the infamous kiwis, teeter on the brink of extinction as introduced mammals ravage their nests and young. European red deer mow down native vegetation, and invasive Eurasian plant species move in at the exclusion of native species. Plantations of fast-growing northern tree species take the place of native beech, rata, and pine. When I told one colleague I was going to New Zealand, he answered, "Why go there? The ecosystem is ruined." All of these changes beg the question: What would the New Zealand landscape be without those things that reflect its essence? What is any landscape without those things?

The bird outside my window is a weka. It is particularly opportunistic among the native flightless birds of New Zealand. It feeds on a variety of foods, including seeds, nuts, fruits, and invertebrates. It can produce multiple broods in a year. Its young are precocial, and can walk just minutes after hatching. Its very presence here around the hut (and that of all the rodents that frequent it) indicates that, unlike the more specialized kiwi, the weka may weather the storm of disturbances that have come with the increasing human traffic to, from, and around New Zealand over the last few hundred years.

Rain pelts the steel roof of the hut and I feel as if I am stuck inside a giant percussion instrument. The weka darts back into the bush and disappears. The gray of the day fades almost imperceptibly to dusk as I prepare soup, then supper, and scratch notes in my journal. As I fall asleep to the sounds of rain, I am comforted by the thought that a hundred years from now there may still be a weka that frequents the tiny green hut called Park Morpeth.

Whitehorn Pass

# IV

There is nothing easy about tramping through the bush on seldom-trodden routes in the Southern Alps of New Zealand. Every one of the varied types of terrain has its own distinct kind of difficulty. Streamside tracks through overgrown beech forests disappear on high-angle embankments as silt-laden floodwaters carry away the soil beneath you. You grab rocks, trees, roots, anything to keep from sliding off and into the water, and finally crawl along through the mud and leaf litter, on hands and knees through the bush as each component of the multilayered canopy drops its load of rainwater down your back. The rivers themselves rise frequently and become impassible, and when a Kiwi (a person, not a bird) tells you to pack extra food for a tramp, it is with this in mind: many a walker has been stranded at a hut while waiting for a river to go down. Others are not so lucky. They step in to ford a swiftly moving stream and are washed away. Some step out onto undercut embankments and the land gives way. In an instant, soil, stone, and person wash downstream.

Above the forests and main mountain rivers there is the subalpine scrub. Here you may find yourself crawling through a tree tunnel of grass trees, cursing your ice ax as its shaft catches on every branch of every shrub, or you may negotiate head-high tussocks and dagger-like Spaniards (Spaniards are a type of leaf-succulent plant with a dagger-like flower stalk) as you step into a hole three feet deep and realize the ankle-breaking nature of the ground's surface. Everywhere water is heard but cannot be seen, as streams that have cut deep into the soil lie hidden by deceptive overhangs of thick grasses. In places, the tangled tussock gives way to boggy ground, where bouncy sphagnum holds your weight with one step, but with the next you are sucked thigh-deep into a black mess of sulfur-smelling muck.

Higher still, the tussock gives way to herb fields composed of a vast array of alpine plant species. Here, finally, the walking is easy for a while, but the herb fields are the exception and not the norm. Too soon you are in a rocky place, steep and rugged, where every graywacke handhold breaks off in your clenched fist. How such rock can be vertical inspires wonder and is indicative of a land that is rising so fast that not even the absurd amounts of water that work away at it can keep the land down. You negotiate your way around dozens of hanging waterfalls, over or through steep-walled sudden gorges, heading for snow that seems indistinguishable from cloud. You kick steps and for once are thankful for the rain because it has created a soft layer of snow to walk on. Rockfall sounds from above, but you can't see it because the mist is closing in around you. And a headwind blasts away at thirty-five miles an hour.

When I awaken to the new day, I immediately notice that the rain has stopped. I enjoy a long breakfast, and over tea I am visited by the now familiar weka. It is late morning before I begin walking.

I follow the upper Wilberforce River for a mile or so before ascending a steep, grassy grade topped by scree slopes. The ascent is well over a thousand feet and tops out on the alpine plateau known as Browning Pass. The top of the plateau is the largest piece of flat ground I've seen in these mountains, and it is filled with the waters of Browning Lake. The scene strikes me as odd. An alpine lake is not an uncommon sight in the mountains, but this one is perched on the highest piece of ground around. Water flows out of it but does not appear to flow in. I contemplate the possibility that this perched plateau (which is really a shallow basin) is some kind of small caldera—an old magma chamber that has collapsed in on itself—but the surrounding rocks are of seafloor origin and do not support the caldera idea.

My attention is diverted by the herb field adjacent to the lake. It is dominated by graminoids—grasslike vegetation—and an abundance of daisies of the genus *Celmesia*. The composite blossoms seem fresh in the cool, moist breeze, as if they have only recently opened to the summer sky. Their specific identity eludes me for the time being. Considering the diversity of alpine species in New Zealand, this is excusable. Of the two thousand species of native flowering plants on the islands, over half of them are alpine plants. In the genus *Celmesia* alone there are over sixty species, fifty of which are alpine. I resolve to take a sample and investigate this matter further in the evening, with my field guide in hand. I take a couple of photographs of the herb field, and scratch a few notes in my journal for reference.

While contemplating the daisies, I notice among them several familiar alpine gentians. They are fewer in number, but there are many of them nonetheless. This intrigues me, because although they are a different species, they are in the same genus and look almost identical to the alpine gentians that grow in the Northern Hemisphere. In fact, both the daisies and the gentians belong to plant families (*Asteraceae* and *Gentianaceae*, respectively) that straddle the equator, and have representative species in both the North and the South. In a place so definitively southern as New Zealand, and after spending time in the distinctively southern forests below, it is comforting to rise up into the alpine zone and see familiar flowers. The gentians, like the daisies, have only just recently bloomed, and this comes as no surprise, because gentians as a group are late bloomers, no matter where they occur in the world. In fact, the gentians of Browning Pass are likely among the first gentians of the season to bloom in this area.

I descend from the plateau along a steep, eroded drainage that soon becomes tangled in thick subalpine vegetation. The drainage leads down to the deep and dramatic valley of the Arahura River, whose waters are conspicuously clear and sapphire blue, a noticeable change from the debris-laden floodwaters I have become accustomed to. The rain has stopped, either because I have traveled far enough to the west or because the easterly storms have subsided, and the rivers are finally going down.

Once I reach the main channel of the Arahura, I follow it down about a mile until I find an obvious track on the west bank. On the track I make good time and travel through the montane woods swiftly. But these woods are noticeably different from those of the Waimakariri River Valley. Here, beech trees are present, but only one of many constituents in the forest. They share space with podocarps such as matai and kahikatea, and grass trees. The feeling of the woods is luxuriant, somehow subtropical despite the nearly subalpine eleva-tion. As I walk on, I notice a clucking sound in the canopy that stops me in my tracks. Above me, a large, purplish-black iridescent bird scuffles about nervously in the boughs of a podocarp tree. It turns, so that for a moment I can see directly into its red eye. Then it launches from the branch, flies by me, through the canopy, and disappears into the trees. It is my first encounter with the New Zealand pigeon.

I continue along the well-worn track at a brisk pace, and after a few more miles I turn due west and pass through the Styx Saddle. The broad saddle forms a watershed divide between the Arahura River and the Styx, both of which eventually drain west and into the Tasman Sea. The saddle is a strange place, unlike any I have yet seen. It is a treeless expanse of tussock, bog, and swamp, with small streams hidden throughout by overhanging vegetation. The saddle ends as the land falls away to the west, and the upland waters reveal themselves in the gushing Styx River. A mile downstream, where the waters of the Styx spread out over a floodplain called Grassy Flats, I realize that for the first time in days I have not been perpetually bathed by rain-water, and I strip down for a cold swim. Nearby, tucked back away from the river, is the hut that will be my home for the night.

In the morning, the sun and unmistakably blue skies portend a truly summer day. The track out the Styx is the most well maintained I have yet seen in New Zealand, but even it is destroyed in many places by huge slips where the steep terrain has given in to the overwhelming forces of gravity and slid into the riverbed. The rock is different here. Instead of graywacke, it is shiny schist, layered along loose foliations. When the land slides, it does so along the planes of the foliated schist, and the rock slides like stacks of shingles on a too-steep roof pitch. Many of the schist slides seem new this year. I navigate amongst freshly broken trees, undercut banks, loose rock, gaping holes, and stinging nettles. But the slides are only perforations in an otherwise continuous, dense, multilayered, complex, temperate rainforest.

Being in the rainforest in the golden sun is like being in the desert during a dark and heavy storm. Something about being in a land-scape in the conditions the very opposite of those that define it brings out secrets that normally remain hidden. As rain brings out new smells and sounds to the desert, so sunlight illuminates the hidden corners of the rainforest. Raindrops clinging to foliage are transformed into twinkling, hanging jewels, silver, golden, and crystal. Innumerable streams and waterfalls glisten and sparkle. As the wind blows across the water's surface, it shimmers and dances like a thousand glass beads strewn across a smooth surface. The color green turns into a hundred subtle hues and tones. The smells, once musty, become fresh, heavily laden with moisture and sweet summer nectar. The sun-warmed air moving over flowers becomes an olfactory equivalent of birdsongs. Mosses and hanging lichens become luminous veils hiding the thick, varied textures of the trunks of trees. The mountains all around become visible, and the forest becomes part of a greater, more com-prehensive landscape. The sky transforms itself from a low, forbid-ding roof that absorbs both sound and thought to a brilliant azure suggestion that beyond the emeralds and kelly greens of tree leaves is limitlessness.

It is difficult to decide where the interior forests end and the rain-forest begins, but it is not difficult to tell that here, on the west slope of the Southern Alps, where the bulk of the prevailing weather sys-tems drop their moisture, is undoubtedly a rainforest. The track is a shallow creek, and indeed in places it appears as if the entire forest floor is covered with moving water. Above and around me are the ancient tree ferns of Gondwanaland, thirty feet tall, forming an inte-rior canopy along with lancewood, rata, and kamahi. Higher up are the boughs of Rimu (red pines) and kahikatea, forming the upper-most canopy of this structurally complex forest. Birdsongs float through the interior of the forest, auditory expressions of the elaborate and beautiful nature of this place.

I have not seen a soul for days. Whether the rain kept others away or I chose an obscure route I cannot say, but the opportunity to experi-ence the mountain landscape of the Southern Alps with the peaceful purposefulness of solitude has been more than I could have hoped for. Something about being alone in a new place allows our thoughts and imaginations to roam wild and free, allows us to attain in our psyche a state similar to the landscape we engage, and perhaps allows us to begin to understand the essence of a place. As I prepare to walk out of the bush, I reflect on water, cloud, crumbled stone, snow, flowers, wet forests, keas, wekas, pigeons, and tree ferns, and try to inhale all of these images into my consciousness as I inhale air into my lungs. As I walk, I try to distill the essence of

Tree fern

this place into words. In some way, the land is like a poem, and the better we understand it the fewer words we need to express it. The fewer words we need to express it, the better we can remember it.

Park Morpeth Hut

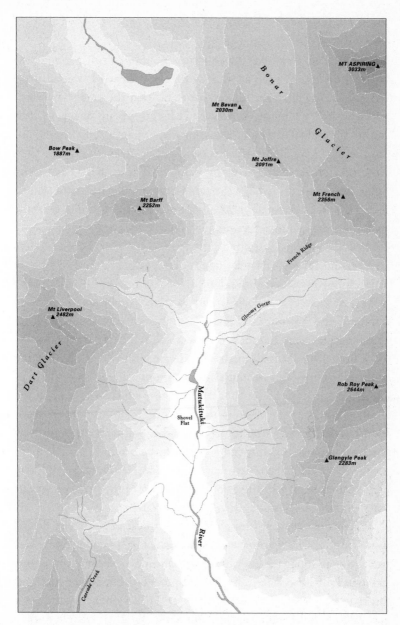

The Mount Aspiring–Dart Glacier region

# V

From the beginning I was drawn to the Mount Aspiring region of the Southern Alps. As soon as I knew the place existed I wanted to go there. I can't explain the lure except to say that the name "Aspiring" probably had something to do with it. It conjures up images of crystalline pyramidal summits reaching triumphantly into the cloud-streaked sky, the alpine wind tossing glittering snow crystals from the peak into ephemeral, wispy banners. The names "Cook" and "Earnslaw" just don't have the same magical ring to them. When Erin Lotz and I planned on meeting up in New Zealand to do some exploring and mountaineering, it was to the Aspiring region that we immediately turned our attention.

Erin and I have been a dating for over a year, and in that time we have climbed together throughout the American Southwest and the Sierra Nevada of California, and explored mountain regions of Scotland and Scandinavia. She is a sophisticated climber, and her experience on technical rock routes exceeds mine by nearly a decade. Back home, she leads all the hard, delicate pitches and I lead all the nasty off-widths and runout chimney grunts. If the truth be told, my forte is really the grueling approaches, which she rues. In the high mountains, she masters the systems, I monitor the pace. It works out. We get around. Here in the Southern Alps of New Zealand, our plan is to spend a week on the ground just getting to know the place, then, weather permitting, attempt a multiday project on Mount Aspiring. For the first part of the plan, we set out for the Dart Valley in Aspiring National Park.

We weave our way through the open forest of Southern Beech, away from our camp on the gurgling banks of Cascade Creek, and out into an open grassy meadow. Sandflies swarm around our exposed limbs and faces. They crawl up my buttoned shirt cuffs and chew at my wrists. We swat at them and occasionally smash them into bloody black pulps against our skin. But the example doesn't seem to deter the other flies, and they keep coming. It is morning, and the wind hasn't picked up yet. The flies still form a discernible layer, a swirling, scribbling cloud for several feet above the ground. I breathe slowly, careful not to suck any into my mouth and nose.

The floor of the West Matukituki River is flat and steep-walled, carved by Pleistocene glaciers that disappeared ten or twelve or fourteen thousand years ago. The valley's profile forms a discernible north-south-trending U, open to the south, ridges rising abruptly to the east and west, on either side of us. At the head of the valley, the terrain rises to the high pass of the Matuktuki Saddle. Framed in the saddle, slightly off to the east, is the massive pyramid of Mount Aspiring. It is by far the highest and most prominent feature in the greater landscape. From our perspective, it is difficult to tell just how high it is relative to the valley bottom. It just looks big, hulking, distant, unknowable. It is indisputably the monarch of the valley.

We pass the large Aspiring Hut and begin making our way up the rough-cut track that leads up the west wall of the valley, toward a pass called Cascade Saddle that offers access into the more remote Dart Valley. The Cascade Saddle route is notoriously difficult because it ascends over thirty-six hundred feet in less than two miles, and we convince ourselves this will be good training. As soon as we start up, we lose the flies and gain a slight breeze. The track is steep, imperfect, and holds our attention. It quickly climbs through beautiful open beech forest to the bush line and emerges out onto alpine tussock grass. As we climb, I realize that the slope angle is well beyond the angle of repose, and a fall, especially if the tussock grass is wet, would send one sliding downslope at rocket speed. We make our steps carefully. The views out at the surrounding mountain landscape get better and better, but we determine not to stop until we reach the saddle, where we are certain they will be even more amazing.

The high point above the saddle is marked by a pylon. Here we stop, drop our packs, and have a look around. The weather is Sierra Nevadan—not a cloud in the sky, cool, dry, and breezy. The surrounding mountains are as clear as belltones, the glaciers crisp, distant valleys soft and blue. Most striking is the new view down into the upper Dart Valley. Here, the meandering form of the Dart glaciers snakes its way down until its snout becomes buried in debris. Above, a score of hanging glaciers cling to the rock of an elaborate cirque, crowned by a dozen summits. In times past, these hanging glaciers flowed directly into the Dart Glacier, feeding it with ice and filling the cirque to the brim. Today, the cirque is almost empty, the upper glaciers cut off and suspended, the Dart shrunken, its nose in the dust. Its current degraded state closely parallels that of mountain glaciers throughout the Northern Hemisphere, and indeed, all of the climatic fluctuations that have affected the ice in the North equally plagues the glaciers of the South. Just on the other side of the saddle, we look down at the valley of the West Matukituki. It forms an almost perfect mirror image of the Dart Valley, separated neatly by the line of the saddle. But no ice remains in the West Matukituki, and its emptiness heralds the fate of the Dart.

We make our way down to the saddle. Along the way, we cross over several exposed outcrops of schist, evidence of recent slips. The metallic sheen of the schist shingles is so intense that walking across them is like walking over aluminum foil, and I can feel the reflected rays of the sun cooking the backs of my legs. The abundance of such outcrops gets me thinking that such slips are common here. As I look around, I am struck by the fact that the entire landscape as far as the eye can see, from the smallest shingle of schist to the most prominent ridgeline, is tilted at a twenty-five-degree angle to the west-southwest. The whole place is in various stages of slipping, like a thick stack of loose paper tilted up. As we follow the track downs-lope, the silvery schist shingles degrade into silvery sand and mica-rich mud, and pulverized silt drifting down mountain streams and reflecting different hues of gray, green, and blue.

We draw closer and closer to the Dart Glacier and soon leave the track and make for the ice. We slide down loose, silvery scree and begin searching for the snout. The valley bottom looks like a wet, muddy gravel pit. This far down the dying glacier, the ice is almost completely covered by a thick mantle of debris. Only in a few places does the ice reveal its underlying presence in desperate-looking cliffs and worn-out towers of exposed ice. Below these is a twenty-foot-high escarpment of ice that looks like it is covered in soot. The upper edge of the escarpment forms a lip that has been undercut and is slowly collapsing into huge, blackened chunks that fall to the soggy mud below. From somewhere deep within this dismal hollow, a milky-gray flood of meltwater rushes out of the bowels of the ice. We have found the terminus of the Dart Glacier.

Below the snout of the Dart, the whole upper valley consists of evidence of the glacier's former size and recent demise. Silver sand and mud from recently pulverized schist lines the banks and covers the beds of the continuously changing braided outwash stream. Till litters the valley in the forms of lateral moraines and long eskers dissected by the changing stream course into isolated heaps of gravel. As we continue down the recently born Dart River, small tarns begin to appear where dead ice chunks have depressed and melted, forming kettle lakes in the valley bottom. They look like translucent blue-gray jewels set perfectly amid the soft silver sheen of the glacial sediments.

Above us, the Marshall Glacier has been calving off icebergs throughout the afternoon, sending refrigerator-size blocks crashing down the thousand-foot cliff face above us. Thankfully, we are on the opposite side of the river, and though the sounds of the ice smashing into the valley bottom are nerve-racking, we pay them no heed other than respect. Just downvalley, the Hesse Glacier hangs still lower into the valley, like a lolling tongue of cracked glacial ice. The valley floor, hundreds of feet below, is a graveyard of melting seracs, recently dislodged from the ice above.

We round the bend to where the Dart River begins its first southerly flow and begin looking for a smooth terrace to make camp on. Nestled on the back side of a lateral moraine, set against the converging folds of the Dart and Snowy Creek valleys, we find a beautiful, small, jade-green tarn lined with silver sand beaches and spongy carpets of yellow, green, and gray mosses. We have reached postglacial paradise. We unload and make camp for the night.

In the morning, I leave camp solo and head up and out of the valley, across the multicolored moss-carpeted shoulder of nearby Mount Ansted. The terrain rises steeply, quickly turning into coarse, thick, tall tussock land broken by deep, silvery slips of schist. The way is hard and long, and I alternate between stepping up clattering, downward-sloping shingles of stone and negotiating tussock vegetation against its grain. The shingles seem to slide and skate a foot downslope for every thirteen inches I gain. Hidden Spaniards send miniature spears of vegetation through my pants and puncture the skin of my legs. I grunt my way upward for two hours, cursing myself for choosing such a stupid route.

I attain the high, rocky point visible from camp, which I assumed was Ansted, but it is not. It is just a low shoulder of the mountain. A long, serrated ridge of schist continues east to the peak. I follow it, fighting the downslope slide of schist shingles as I make the traverse. Eventually, I reach less-steep terrain and continue up snowfield after shingle slope after snowfield after shingle slope. Ansted draws near as the rock becomes steeper, and I ascend the sliding, clinking schist to the top. Once again, I am struck by the fact that these mountains, supposedly still in the midst of their genesis, seem to be falling apart.

There are plenty of places on the planet where geologic forces are at work at a quickened pace, but New Zealand is in a category all its own. These mountains were thrown up so hastily they can't hold together. The schists and graywackes that make up the range are so soft you can wear them down with your fingertips, yet still, amidst the onslaught of the elements, the mountains are rising fast. To get an idea of how fast all this is happening, the summit rocks of Mount Cook, at well over twelve thousand feet the highest point in New Zealand, were beneath the sea less than a million years ago.

Once-adjacent rocks on either side of the Southern Alps have moved laterally up to three hundred miles. The North Island was home to the most powerful volcanic eruption of the last five thousand years.

The bulk of what is today New Zealand began as sea floor sediments that washed off the supercontinent Gondwanaland. Two hundred million years ago, ancient rivers delivered debris to the sea, and the sediments settled in extensive deltas, much like what is happening at the mouths of the Nile, Amazon, Mississippi, Yukon, and McKenzie Rivers today. These sediments stacked up and accumulated to depths thousands of feet thick.

When Gondwanaland began to break up, the seafloors offshore from the continental landmasses began to subduct beneath the continents. In the midst of the subduction zone, the leading edge of the continent acted like a bulldozer, scraping up the thick sequences of seafloor sediments into tilted wedges. As the continent pushed over the subducting seafloor, these wedges of sediments were squeezed, metamorphosed, and pasted onto the leading edge of the continent. Over millions of years, the amount of this newly accreted material became considerable, extending the edge of the still-connected landmasses of Australia and Antarctica.

As Gondwanaland continued to split apart, new spreading centers opened up. One of these formed between Australia-Antarctica and the newly formed rocks of New Zealand, driving a wedge between New Zealand and its mother continent and rifting them apart. Whatever life was present on the dry land of New Zealand was cut off, isolated from the continental mainland forever. The Tasman Sea was born, and by eighty-five million years ago, New Zealand had become an independent continental fragment, a microcontinent adrift, an outlier at a disjunct edge of the Australian Plate.

Over tens of millions of years, the rock of New Zealand slowly eroded away and washed out to sea. As the ground wore down, seawater flooded the microcontinent, and only the highest points remained dry land. Eventually, by twenty-five million years ago,

New Zealand, at the outer edge of the Australian Plate, butted up against the expanding Pacific Plate, and a new collision zone formed. To the north, the Pacific struck New Zealand at a direct angle, and subduction of the oceanic crust began to form the arc volcanoes of the North Island. To the south, the Pacific struck New Zealand at an oblique angle, twisting and buckling the crust along the boundary to form the Southern Alps. Today, New Zealand straddles the plate boundary of Australia and the Pacific, and the tension between the two is borne out in the characteristic mountains of the North and the South. The islands themselves are but one small link in the great arc of mountains that describes the Pacific Rim. But among all of the volcanoes, earthquake-ridden zones, and stacked mountains that surround the Pacific, the mountains of New Zealand are among the most active. On a tectonic map, where black dots represent recent volcanic and tectonic activity, the islands of New Zealand are covered with so many dots they appear solid black.

The summit is false. It is only the beginning of a high, exposed, sawtooth ridge of loose, sliding schist that extends east to the true summit, a knife-edged arête of some of the most nightmarish rock imaginable. I pick my way along the ridge very slowly and carefully, continuously reminded of my precarious situation by the shifting of rock beneath my hands and feet and the yawning bergschrund, the uppermost crevasse of the glacier immediately to the south of the arête. I yearn for Sierran granite. Several times I am forced to go down and around impassible gendarmes. The final climb to the summit means trusting my life to a series of high-angle schist plates that are only loosely attached to the summit blocks, ready to slide along their bedding planes at the slightest provocation. I move slowly, deliberately, delicately. Luck helps me to the top.

The view all around is of metallic, tinfoil mountains of bedded schist, slopes streaked with silver and the green of alpine vegetation, peaks jutting out like pointed hogs' backs amid fields of white and blue glacial ice. It is astounding, a bird's-eye view of the accumulated, tilted, worn-down, and presently rejuvenated, recycled crust of Gondwanaland. Is it worth it? Looking across the broken ridge I must recross to get home, I wonder. Tiny bits of schist slide over each other as I run my fingers over the rock of the summit block. Away across the Dart Valley, the Marshall Glacier sends a block of ice the size of a small car careering down a water-streaked cliff. I fantasize about hopping on a shingle and riding it all the way back to camp, four thousand feet below.

Mountain terrain

# VI

Mount Aspiring is called the Matterhorn of the South because, like its northern counterpart, it is a classically sculpted, symmetrical glacial horn that rises as a singular mountain, head and shoulders above anything else around. Although it is not the highest mountain in the range, it is arguably the most beautiful, the most captivating, and the most inspiring of imagination. Together with its surrounding glacial plateau, it forms a nival island of glaciers and stone, a crystalline pedestal culminating in a pyramidal peak of immense proportions and elegant lines.

Shafts of evening light pour through Cascade Saddle and illuminate the mountain massifs all around us. Quiet cumulus clouds, delicately entangled around the summit rocks, slowly release their hold, relax, and dissipate into the coming night. Glaciers take on hints of color not seen by the light of day, each lasting only moments before melting into another: silver, gold, amber, peach, rose, gray. We await the luminescence of the moon and the light it will cast on the mountain landscape. I have seen this progression a hundred times on a hundred mountains, and it always makes me stop and think, makes me forget about the things I need to forget and remember the things I need to remember.

We are camped on the French Ridge, the southwestern buttress of the great plateau upon which Mount Aspiring sits. We worked hard to get here, all the way up through the lichen-hung forests and flower-flecked meadows of the West Matukituki Valley, up the lower French Ridge, which rises three thousand feet in a mile through gnarled, twisted beech trees and roots, subalpine scrub, and tussock grasses, and continues all the way to the edge of the plateau ice. After our climb, we stopped at the nearby French Ridge hut to visit with the warden and have tea. As evening set in and other parties began to drift into the hut noisily, we said good night to the warden and made our way up the ridge a ways to make camp.

The moon rises full, and we lie awake talking about the upcoming climb. What will the climb up to the plateau be like? Will there be a bergschrund where the ice of French Ridge meets the ice of the plateau? Will we be able to get around it? What will the glacier be like? Will there be icefalls? How long will it take us to cross from the French Ridge, on the southwest edge of the plateau, to the Shipowner Ridge, on the northwest side? What will the climb up Aspiring be like? Will there be other parties? Will we need to travel roped the whole way? After close inspection of our map, we have answers for some of these questions, but for others, only time on the ground will tell. It is past eleven when finally we rest.

There is a noise outside that sound like fabric tearing, and in the moonlight the shadow of a kea is cast on our tent. I unzip the fly and find the culprit. He is dancing around on the rocks outside the tent in the full moonlight. He looks thrilled, and wants to play. Looking down at the new four-inch rip in my tent fly, I scold him bitterly. He hops off a ways from the tent. I zip the flap and settle back in, but within a minute he is back, poking at the fly. Erin sinks deeper into her sleeping bag in terror. I watch as the kea pokes its hooked upper mandible through the fly and tears a neat, foot-long incision in it. I just stare in disbelief. Keas never rest when the moon is full. He will ransack our tent all night long. We have no choice but to pack up and move inside the hut.

Midnight. The last climbing party stumbles wearily into the hut and sets about unpacking, clinking metal equipment together, crinkling food wrappers, and cooking supper in apparent disregard for the seven others packed helplessly into the adjacent bunk room. I dread the night.

One o'clock. I lie awake. Each time slumber overtakes me, the climbers in the hall open or shut the door and it wakes me up. I curse the kea, but know that they would not behave as they did if not for the continual presence of humans here. I curse the climbers. I curse the hut and all huts. I yearn for the wild mountains of western North America, for wilderness camping, for freedom. One by one, the climbers come into the bunkroom, and each of them falls asleep instantly. Within minutes of his arrival, one of them starts to snore, and I listen in disbelief that someone with so much mucous rattling in the back of his throat could even be alive. After some minutes of this, I get up the nerve to reach down and shake his feet. He stops snoring long enough for me to settle back in and put my earplugs in, then he starts again. The earplugs are of no avail. The entire room is vibrating. Next to me, Erin sleeps soundly.

Two o'clock. I give up. I grab my bedroll and move out to the hallway. There, I crawl back into my sleeping bag. Even with my earplugs in I can hear his snoring through the walls. I vow never to sleep in a hut again.

Three thirty. After an hour of half-sleep I am awakened by a party of two who are anxious to begin their climb while snow conditions are ideal. They take over an hour to leave, making all possible noise in the process with no detectable attempts at being considerate. They walk by me at least a dozen times. There is no God. If there were, he would stop this.

Five thirty. The warden awakes and goes into the kitchen. I haven't slept since the three-thirty team woke me up, so what's the difference. No rest for the weary. Sleep is for the dead.

Six thirty. Erin comes in to wake me up. She doesn't need to, because I've been awake all night contemplating bailing off this God-forsaken ridge and going somewhere, anywhere, where I can get some rest. The idea of climbing makes me sick to my stomach. In the bunkroom, the same guy is still snoring away, sleeping like a baby. I secretly want to go in there and scream in his ear for six hours. I give up completely and get up.

French Ridge spreads out before us in the morning light, a stairway of rock and ice to the high plateau of the Bonar Glacier. We walk up and into the light, trading rock for clean snow as sweat beads on our brows and trickles down our faces. I walk in sneakers, light- and fleet-footed, and we briskly make our way up the snow ramp as the slope angle increases and crevasses begin to interrupt the continuity of the snow surface. We scamper up to the rocks at the edge of the ice and stop to switch into our boots and crampons. Below us is the deep, gray chasm of Gloomy Gorge, home of a dozen hanging waterfalls. Continuing, we come to a gaping crevasse at the Quarterdeck Pass, where the French Ridge joins the main plateau. A three-foot-wide swath of snow allows for safe passage around the right side of the crevasse. As we ascend it, we hope it will still be there two days from now, on our return.

The Bonar Glacier is stunning: wide, beautiful, and undulating in graceful curves. It seems around twenty square miles, not including the many cirque glaciers feeding into it. But however captivating the glacier may be, Mount Aspiring, looming four thousand feet above the plateau immediately in front of us, grabs our attention. It is a huge and intimidating mountain. My psyche crumples for a moment, then reassembles itself for the task at hand.

We stop to set up our glacier rigs, rope up, then set out across the girth of the Bonar. The going is easy, the snow surface like crushed velvet, marked only by a few obvious lateral crevasses easily stepped over. As we pass the foot of the Southwest Ridge of Mount Aspiring, the glacier convulses for the first time, but we pass by easily alongside seracs, caverns, and folded, swirling ice. As the glacier bends down to the final sweep of ice before the Shipowner and Northwest ridges, it becomes increasingly broken up, and the ice is bare, blue, crunchy, and in places dirty. We carefully wend our way through a maze of open crevasses, zigzagging toward the nearby ridge.

It is midafternoon when we step onto the terra firma of the Shipowner Ridge. The thought crosses my mind that we could attempt the summit today. We wouldn't see a single party this late in the day, we would enjoy the summit to ourselves in the long light of evening, and we would recross the Bonar Glacier tomorrow morning, while the snow is still hard. I suggest it to Erin, but she wants nothing to do with such a plan. Her reasons are sound. It looks like the weather will be just as clear tomorrow as today. Climbing in the morning will give us better snow conditions on the peak, where we need them most. I am tired and could use some rest. Why rush? We look for camp.

The Colin Todd hut is perched before us on the rock of the ridge, but after last night's hut experience we carefully avoid it. We remain out in the light amid the low, shifting cumulus clouds, and seek a campsite far enough from the hut that local keas are less likely to harass us. Within minutes we find such a spot, a flat perch over-looking the vast expanse of the Bonar plateau. There we spread out in the sun and rest as refreshing breezes waft over the ridge from out of the north. Aspiring looms large against the eastern skyline, a reminder of what is to come.

We awake at 4:00 A.M., slowly, uncertainly, half willingly. It is dark without a suggestion of dawn. Day is a distant hope and the sun a far-off light. Up the ridge, six lights twinkle like a linear constellation as a group of climbers slowly ascends the flank of the Northwest Ridge.

Erin wants to get up and go. I admit that the sight of other climbers is an incentive to get moving, but I advocate for another hour of rest. We are in no hurry, and will move swiftly once we start. Our strategy is efficiency: a small team carrying minimal weight, traveling unroped, moving at a quick pace. The early morning habits of other climbers are no reason to spoil a good night's rest.

An hour and a half later, at 5:30 A.M., a pale gray light hovers over the ridge. Our bellies are full of hot oats and tea and cold mountain water, and we are laden with snacks, water, extra sweaters and parkas, boots, crampons, and one ice ax each; we set out up the dark silhouette of the ridge. We skip across hard schist folded and corrugated like tin sheet metal roofing, polished and rounded into bulbous humps, elbows and shoulders of the long western arm of Aspiring. We pass shivering edelweiss flowers with feltlike white petals wet with dew and delicate feathery-leaved buttercups nestled quietly in the rocks. As the dawn waxes, we see that the flanks of the mountain, invisible to us at first, are in fact wreathed in cloud. As we ascend, we rise into the gray mist from beneath, and the depth of the world goes flat and dim.

We ascend to a small peaklet marked by elevation 2,151 meters, the point where the lower Shipowner Ridge connects with the upper Northwest Ridge. Here we encounter our first navigational uncertainty. We split up and scout around either side of the peaklet, but both routes lead to precipitous drops onto barely visible glacial ice below. To the northeast of the ridge we find smaller drops in the rock, but the downclimbing looks hard and uncertain in the gloom. Finally, after half an hour of deliberation, we find a sketchy line that traverses the rock of the peaklet's north face. We follow it across a pitch of loose, unpredictable rock, making some tricky balancing moves and passing packs back and forth, and finally stepping down onto a settled pile of shingles well below the ridge. We are relieved to see an easier route back up, and follow it around to the other side of the peaklet. As we climb, I am reminded of my scramble on Mount Ansted several days ago and am thankful for the relatively firm rock of Aspiring. It is some of the best rock I have seen in the Southern Alps, and it comes as no surprise that the most resistant metamorphic core is left standing as the highest mountain.

The clouds thin long enough for us to see the long, spinelike sweep of the ridge that leads to the upper northwest buttress. It looks like the fine line of a woman's back. We stand poised on her tailbone, and before us the small of her back dips down slighty before rising elegantly to her shoulders, a clean line broken only by a single conspicuous gendarme before the base of the buttress. Beyond the gendarme, the ridge is wrapped in mist. Within minutes the mist lowers, and the mountain is once again wrapped in a veil of clouds.

We move quickly across the sleeping beauty's back, as if not to awaken her lest she shake us off into the swirling abyss. When we reach the first gendarme, we pass easily to our left, briefly leaving the top of the ridge. The clouds thicken to a bright, flat white in the full morning light. Somewhere ahead looms the northwest buttress, but we have lost it in the mist.

As we continue, the curve of the ridge bends upward, and a second gendarme materializes from out of the mist, a precursor of the formidable, invisible buttress that lies beyond. The path of least resistance leads naturally to the right of this obstacle, and we climb over rock ledges and snow to regain the ridge. Once back on it, for the first time we come face to face with the intimidating buttress. Silently, we both measure our skill and experience against the dark, dripping stone looming immediately before us. After some deliberation, we follow a broken cairn that leads to the left, north of the ridge. We neither ascend nor descend, but rather traverse ledges of gravel and broken rock, past a shrinking snowfield gurgling with meltwater. The further we go, the more the task of regaining the ridge grows into an impossibility. But the tinkling sounds of running water are tantalizing to us, a siren song in our ears. On this cold morning when no snow has yet begun to melt, fate has brought us to the only flowing water on the mountain. We huddle in a sheltered alcove, where tiny streamlets of cold, sweet water are decorated with buttercups and *Celmesia*, and drink our fill.

After a short rest, the wind begins to bite at us, and rather than put on more clothes we decide to keep moving. We backtrack across the snowfield and find a steep route up wet ledges that leads us back to the main ridge of the buttress. Climbing feels good, and we move with practiced fluidity. Once back on top, we settle into a steep slog as bitter winds batter us from various directions. Ahead of us in the rushing fog, I make out the vague forms of people sitting in the snow. As we draw nearer, their outlines become clearer, and we recognize them as the party of six we saw from below in the wee hours of the morning. They are dressed in full mountaineering regalia from head to toe, and carrying an array of pickets, flukes, screws, cams, nuts, slings, and ropes. They seem surprised to see us, as we appear as if we are out for little more than a morning jog. We greet them with good mornings and hope for visibility on top. We all agree that the climbing is excellent and the day fantastic, even though what lies ahead remains cloaked in swirling clouds and we can barely hear each other in the driving wind. We ask their permission to pass by them, and continue up.

The snow ends after a few hundred feet and gives way to steep rock, which leads to more, even steeper snow up above. I sit down and exchange my sneakers for mountaineering boots. As I look back at Erin coming up the hill, the sky opens up to reveal an endless sea of purple clouds below us. The sea extends to the southwest as far as the eye can see, punctuated only by the twin peaks of Mount Earslaw in the distance, which rise above the clouds like a mythical Olympus. Light pours through this ephemeral opening, and the surrounding mist glows with a soft, bright, white light. As the mist continues to separate, a narrow strip of clear, deep blue appears for mere seconds and is gone. The mist shifts, and as if but a fleeting vision of hope, the scene is swallowed in clouds. But the sky is opening, and we will rise above the weather today and stand on the glimmering snow of uttermost Aspiring, sunlit above the clouds.

We follow the steep, wet rock up to snow. The clouds envelop us, then part, then envelop us again. But they grow wispy as we ascend, as if uncertain of themselves. The frozen snow forms a steep, unbroken ramp that leads up, up to the definite and undoubtedly illuminated summit of Mount Aspiring. For a while we follow a narrow band of rock between the snow and the sharp edge of the mountain, which eventually leads us to the final summit snowfield. It is steep, maybe fifty degrees, several hundred feet to the top. We stop, put on our crampons, and follow the bright snow toward the summit.

The aesthetic of high-mountain summits is unique and incomparable to any other place on earth. There is no substitute. On the summit of Aspiring, the metaphors of the imagination and the empirical realities of the world meet, and the moment seems mythological in scope, refreshingly legendary, ascended from the mysterious dimensions of mist to a new world of brilliant clarity. Somewhere at the ill-defined and ever-changing boundary between such clarity and obscurity, we live our lives in the world. Today we go deep into both extremes.

The top of Aspiring is glowing, humming, singing as we step up the crunchy snow. We climb steadily, methodically, inspired by the ever-growing sight of the summit. It grows in our minds as it grows in our eyes, a perfect snow-covered tip of perfectly excavated horn, fluted on four sides by the work of glaciers that swelled and shrank over ages on its flanks, buttressed by carefully beveled ridges of the hardest stone around, mantled by smooth snow gracefully sculpted by the work of wind, brilliantly lit now by the midmorning sun, a jewel set atop a pyramid floating in a sea of billowing, pastel purple ether. Few mountains share this celestial space between the clouds and the heavens; the stout form of Mount Earnslaw rises to the southwest, and the massive form of Mount Cook looms large to the North. To the east, the clouds break and clear, revealing a succession of green ridges extending like earthen waves out toward the rolling Canterbury Plain. But in all other directions, the land beneath the uttermost peaks is cloaked, shrouded even as it was a thousand years ago, even as it will be a thousand years from now, Aotearoa, the Land of the Long White Cloud.

The Bonar Glacier

Summit tracks

# CROSSING LATITUDES

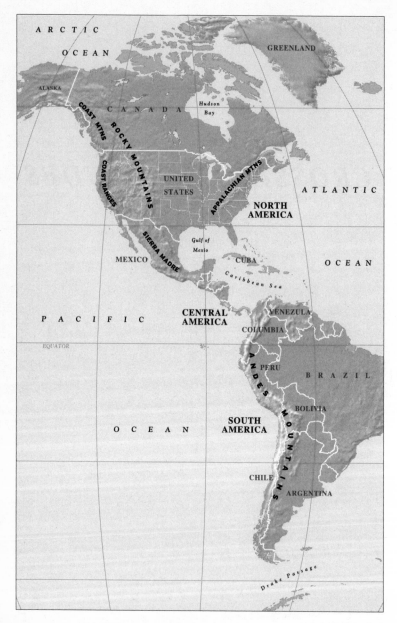

The Americas

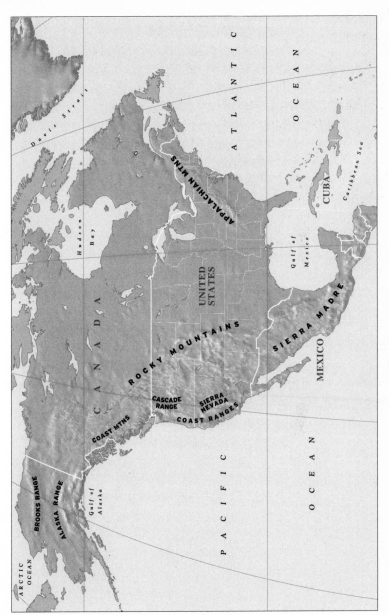

The North American Cordillera

# The North American Cordillera

## I

**CORDILLERA PATAGÓNICA.** Aconcagua. Cordillera Domeyko. Cordilleras Occidental, Central, and Oriental. Cordillera Real. Cordilleras de MÇrida and de la Costa. Cordilleras de Talamanca, de Guanacaste, Chontalena, and Isabella. Sierra Madre Del Sur, Occidental, and Oriental. Sierra de San Pedro Martir. Sierra Juarez. Sierra Nevada. Southern Coast Ranges. Northern Coast Ranges. Klamath-Siskiyou Ranges. The Rocky Mountains. The Cascades. The Olympics. The Pacific Ranges. The Icefield Ranges. The Alaska Range. The Brooks Range. The Aleutian Range. The American Cordillera encompasses all of these ranges and scores more as it sweeps from south to north, across over 120 degrees of latitude, forming a wall of mountains that wraps nearly halfway around the globe. Together with its eastern counterpart in Asia and Oceana, the Cordilleran mountain system describes an immense arc that binds the Pacific Basin. The Cordillera is unrivalled in extent, not even by the whole expanse of the Alpine-Himalayan system. Its length is nearly equal to the circumference of the earth. It is the uncontested most extensive mountain system on the planet.

In short, the entire Pacific Rim Cordillera exists because the continents of Asia, Oceana, and the Americas are closing in on one another, slowly but surely overriding and devouring the Pacific Ocean basin in the process and throwing up mountains all along the boundaries. Interestingly, it is the present spreading of the smaller Atlantic that is largely responsible for both the demise of the Pacific and the development of so many mountains in the collision zone. As the Atlantic basin opens up, it pushes Eurasia and Africa one way and the Americas the other. When crust is made in one place, it is devoured in another. For every inch the Atlantic opens, the Pacific closes an inch. Where the continents on either side of the Atlantic once fit together like a jigsaw puzzle, they are destined to assemble once again on the other side of the earth. In some two hundred million years the Pacific will be shut, and Asia and the Americas will have a reckoning of Himalayan proportions. It won't be the first time such a catastrophic collision has occurred. This has been the dance of the continents for over a billion years. One billion years ago they assembled to form the supercontinent Rodinia. It rifted and spread. Eventually, on the other side of the globe, the continents formed Pangea. It rifted and spread. Currently, the continents are somewhere in the middle. Each time they come together, mountains are made. Each time they rift, those mountains are broken apart, washed away, but new mountains are born at the outer edges of the newly spreading continents, where the continental crust meets that of the surrounding ocean basins. The Pacific Rim Cordillera occurs where the continents that once formed Pangea are spreading and colliding with one of the largest ocean basins the planet has ever known. Thus the extent of the Cordillera.

Even a cursory sweep of the entire Pacific Rim Cordillera would take years to complete. It would involve traveling some twenty thousand miles through a score of countries on four different continents. It is so grand a vision as to be overwhelming. Where to start? In my case, it was close to home. When offered the opportunity to teach a ten-week geography field course focusing on exploration and study of mountain landscapes, it didn't take me long to figure out where to go. I looked for three mountain ranges that were different enough to allow for rich comparative study, but still sharing a few core characteristics. They had to be relatively close together to minimize travel expenses, but far enough that their differences would be more than subtleties. The map of the North American Cordillera imprinted itself on my brain and would not go away.

The mountains spanning the western edge of the North American continent straddle over thirty degrees of latitude. Their southernmost extensions reach into subtropical latitudes yet host alpine vegetation one would recognize in tundra ecosystems anywhere in the Northern Hemisphere. These outreaches, such as the Sierra Nevada of California, hold the lasting imprint of the once vast mountain icecaps of the Pleistocene in the very same granitic rock that lies at the foundation of the Cordilleran mountains from Patagonia to the Aleutians. As we move north, the cavities in stone that the glaciers left are gradually filled in, until, having seen the hopeful mountain glaciers of the Cascades, one arrives at the Icefield Ranges of the Alaska-Yukon border country, the holdout of the most extensive nonpolar icefields in the world, a glimpse today of what the mountains to the south looked like ten thousand years ago. As the glaciers grow and overwhelm the mountain landscape, so, too, wildness grows and overwhelms civilization. While the Sierra Nevada and the Cascades are islands of wilderness is seas of civilization, in the North, civilization becomes the island, and wilderness a sea big enough for moose, caribou, grizzlies, and wolverines to roam wild and free. The mountains, like fingers reaching down from the North, have for millions of years been pathways for flora and fauna, glacial ice, and wildness. A quick look at a map reveals this truth. Even a Rand McNally road atlas illustrates conspicuous blotches of green, federally protected wilderness areas, all along the North American Cordillera, especially in the lower forty-eight. Closer inspection reveals small patches of white and blue within those fields of green, and these are the southernmost remaining glaciers of the Cordillera. To see the subtleties of flora and fauna requires real time on the ground. A journey north, crossing latitudes along the North American Cordillera, is a journey to the source of life in the mountains of the Northern Hemisphere. It is a journey into the source of glaciers that shaped the mountains to the south so long ago, and continue to do so in the North. And it is a journey into the last holdout of wild landscape that once covered the continent. From these three perspectives—physical geography, biogeography, and wildness—a trip along the North American Cordillera can't be beat.

The American Cordillera forms roughly half of the greater Pacific Rim Cordillera and is physically separated from the eastern half of the Cordillera by the Pacific Ocean.

The American Cordillera itself is physically subdivided: half of it occurs on the continent of South America, and half on North America. Only in recent years (from a geologic perspective) have the two land-masses even been connected by the thin isthmus of Central America. While the separate continents of North and South America form a natural physical boundary between the Cordilleran mountains to the north and those to the south, the tropics form a biological boundary. The tropics form a swath some three thousand miles from north to south that includes the entire midriff of the Americas, as well as northernmost South America. While this thick belt of climatic con-ditions results in unique mountain flora and fauna all its own, it effectively separates the flora and fauna of the temperate, subpolar, and polar regions of the north and south, isolating them in their respective hemispheres. In the South, where the amount of landmass is small, the geographic isolation of temperate and subpolar flora and fauna has led to some pretty fantastic renditions of life, while the north has become relatively stable and evolutionarily conservative.

Because most of the landmasses on the planet are presently huddled around the Northern Hemisphere, terrestrial life has long had plenty of room to roam, mix, and sort out evolutionarily. The vast expanses of the northern Arctic tundra and subpolar boreal forests form the largest contiguous ecosystem in the world, and have no equivalent in the South. Relative to the small, isolated landmasses at similar lati-tudes in the Southern Hemisphere, in the North competition for resources is high, ecological releases fewer and farther between, inter- and intraspecies relationships better established, and evolutionary processes slower and more refined. Because the landmasses in the North are all either joined or separated by relatively small stretches of ocean, given the same habitat the same species may occur in Alaska, Canada, Maine, Iceland, Scotland, Norway, and Siberia. This holds true particularly for species adapted to cold polar or subpolar climates.

The list of such species is long, and includes mountain sorrel, arctic willow, wolves, caribou, and moose, to name just a few. The list of closely-related species with analogs in different northern areas is even longer, and includes all of the conifer genera, just about the entire heath family and American and European black bears and beavers.

Mountains play a critical role in the distribution of the flora and fauna of the Northern Hemisphere by virtue of the fact that elevation simulates latitude. Simply put, at a given elevation, the further one travels up the mountainside, the colder and wetter the climate gets. As the temperature drops and rain and snow begin to fall, the climate increasingly resembles that of sea level to the north. As climate is the ultimate arbiter of the geography of natural communities, not only does the climate resemble that of the North as you go up a mountain, but so does the flora and fauna. In the Sierra Nevada, at 1,000 feet above sea level you will find yourself in an oak savannah, at 5,000 feet in the thick of a pine-oak forest, at 8,000 feet surrounded by conifers well adapted to subzero temperatures and heavy snows, at 11,500 feet at an alpine treeline that resembles that of the arctic, and at 13,000 feet in tundra that looks like the west coast of Greenland. The story is much the same for the Cascades, except that they are farther north and all the ecosystems are depressed. At 1,000 feet you are already in montane forest. At 5,000 feet the trees are already starting to thin out. At 8,000 feet you are well above treeline, atop hulking glaciers flanked by dark rock outcrops garnished with thin patches of tundra. At 11,500 feet you are well into the nival zone, in which climatic conditions are so rigorous no vascular plants can survive. At 13,000 feet even the remaining lichens are struggling. Still farther north, in the Icefield Ranges of the Alaska-Yukon border country, at a thousand feet you are already approaching the upper limit of the forest. At 5,000 feet you are well into the alpine tundra, and may be on expanses of glacial ice miles wide. At 8,000 feet you are well into the nival zone and are surrounded by ice. At 11,500 feet the lichens drop out and the entire world is cloaked in a mantle of ice a mile thick. Only the uppermost peaks rise out of and above the ice, and these are barren islands of stark stone.

That the Cordilleran mountains run north-south gives the well-developed flora and fauna of the North easy migration pathways, or inroads, to the south. To say that the mountains are sky islands is not enough. They actually form a nearly contiguous sky country of montane/boreal and alpine/tundra habitat, shaped something like Chile (but with a few important gaps) if you were to look at it on a vegetation map. But the farther south the mountains reach, the higher and higher these habitats occur, and eventually, in the Southern Sierra Nevada, the mountains are no longer high enough to support alpine tundra. Even farther south, as the trend continues, they are no longer high enough to support boreal-type forests. By the time the Cordilleran mountains rise once again above treeline (deep into mainland Mexico), they are so far south that the chain has been broken, and few if any species with origins in the North can get there. The Sierra Nevada, then, represents the southernmost extension of a number of such species. From a biogeographic perspective, it is the logical place to begin the journey north.

From the perspective of glaciation, the Sierra Nevada is also the place to begin the journey north. The Sierra Nevada is home to the southernmost glaciers of the North American Cordillera. Once again, south of the Sierra the mountains bow too low to host glacial ice, and it is not until the distant Mexican volcanoes that the mountains once again glimmer white and icy in the sunlight. The few remaining glacierettes of the Sierra Nevada are mere suggestions of a once-great Pleistocene mountain icecap that covered the range and continued north in polka-dot fashion up the chain of volcanoes that are the Southern Cascades, becoming a contiguous icecap again in the North Cascades and extending all the way to Alaska.

The North American Cordillera is unique among mountain systems of the world because it is one of the only extensive high mountain systems that never gave rise to cultures synonymous with the mountain landscape. As such, it is one of the only mountainous regions of the world (aside from the polar regions) that is almost wholly wilderness. While all the other permanently inhabited mountainous continents

(Asia, Europe, South America) have their respective mountain peoples, North America has none. To be sure, various first nations of the Americas lived around and identified with nearby mountains, used resources from the mountains, traded across high passes, and frequently made seasonal camps above treeline, but as yet no one has come up with any compelling evidence to suggest permanent habitation of high mountain environments. Considering the relatively low population of North America prior to European contact, combined with the abundance of natural resources found particularly on the rich coastal slopes of the Cordilleran ranges, the dearth of definitively mountain cultures among first nations is not surprising. From the perspective of the hunter-gatherer or the agriculturalist, high-mountain landscapes are among the most marginal habitats. Only necessity pushes people to develop the technological innovations necessary to survive in such places. Thus, prior to European contact, the high mountains of North America remained wild, uniquely devoid of permanent human presence.

The advent of industrialization that followed European contact changed the cultural geography of western North America (and the whole world, for that matter) forever. The combination of mass-produced goods and improved transportation (the railroad in the case of western North America) ensured that people could depend on nonlocal resources for everything including food, shelter, fuel, and a surplus of nonessentials. In time, this allowed for exponential population growth in the west, and even as urban centers have exploded into the lowlands, Western civilization has, for the first time in human history, made permanent inroads into the mountains of the North American Cordillera. But are these mountain people? Traditional mountain cultures are so defined because to a large extent they derive their resources for living directly from the landscape they inhabit, and so develop a long-standing coevolutionary relationship with landscape that is intrinsic to all aspects of life. Such coevolutionary relationships may be impossible for post–Industrial Revolution cultures, whose relationship with landscape tends to center around aesthetics and recreation, to achieve. At best, such coevolution occurs at a greatly slowed pace. Meanwhile, some of the last wilderness in the world gets butted up against and chewed away by modern American culture's incessant appetite for expansion.

Like glaciers and northern flora and fauna, the wilderness of the North American Cordillera becomes increasingly expansive to the north. While the urban centers of California encroach upon and shed unwanted light on the Sierra Nevada, and the growing metropolitan areas of the Pacific Northwest isolate the Cascades and the Olympics, across the Canadian Border the Cordilleran mountains feel far from the influence of cities. The Icefield Ranges of the Alaska-Yukon border country lie amid a wilderness that spans the continent, where the trappings of modern culture are relatively few and far between. From the perspective of wildness, the Sierra Nevada, in the heart of the most heavily populated state in the most heavily populated and most extensively developed country in the Western Hemisphere, is once again a good place begin the journey north. And somewhere above the sixtieth parallel is a good place to finish.

High Sierra

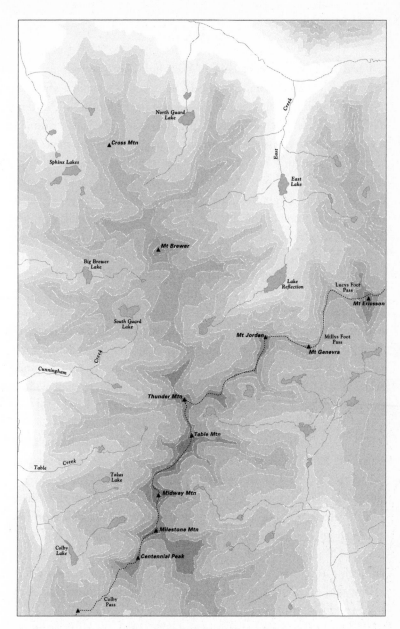

The Great Western Divide region

# II

If I had to describe the essence of the High Sierra Nevada of California in one vivid scene, this would be it: a clear, swollen sapphire lake with a glimmering sheen of surface ripples, brimming with trout, nestled in a wide, glacially polished granitic basin. The lake is rimmed on one side by gleaming snowfields textured with the ridges and troughs of sun cups, and the snow slants up increasingly steeply to steel-gray, water-streaked blocks, exfoliating sheets, and eventually precipitous cliffs of granite, dark against the bright snow, pale against the deep royal blue of the sky. On the other side the lake is rimmed by meadows, dry in places and wet in others, the green of sedges and herbs interrupted by conspicuous bursts of magenta: the flowers of mountaineer's shooting star and bog laurel, and studded with rounded granite boulders fringed by pink-blooming red mountain heather. The meadows slope gently into groves of widely spaced conifers and sweeping, smooth granite slabs. The sounds are of water cascading over stone, the wind whistling through pine boughs, tinkling bird-songs, the whine of mosquitoes. The smells are of nectar, pines warmed by the sun, the sweet freshness of freshly opened willow catkins and leaves. The air is warm but not hot, the wind cool but gentle. It is June, at 10,500 feet.

We have climbed thousands of feet to get here, and it is time for a break. At Sphinx Lakes, in the heart of the southern Sierra Nevada backcountry, we warm ourselves in the sun on the lakeshore after a cold swim. My students are quiet. It is still the first week of the course and the climb up here from the dusty bottom of Kings Canyon has not been an easy one. We are laden with packs containing ten days' worth of food, clothing, shelter, notebooks, a small library, and mountaineering equipment. We left the trail miles ago, two thousand feet below, and are following Sphinx Creek and its paternoster lakes to their uttermost source in the High Sierra.

There are thousands of fish in the lake. They are rainbow trout, and every color in the landscape—gray, silver, blue, pink, green—

seems reflected in them. I watch them, entranced, as they swirl around in the cold, clear water. After a while I start to grow hungry. I slowly stand up and begin making my way over to the lake's outlet. There, I imagine hundreds of fish literally clogging up the spillway, spawning, slow, easy to catch. I have caught half a dozen such fish, each fourteen inches long, in half an hour without difficulty. Though camp is still over a thousand feet up and miles away, the extra weight of a few fish seems worth it. I crawl to the outlet on my belly, using my elbows to pull myself along. The fish are there, but they are wary. Every single one of them scatters. So much for fish tickling today.

From the deep blue lakes we traverse the wide, open alpine basin of upper Sphinx Creek. It is rolling country compared to the gigantic glacial staircase we have been ascending since the floor of Kings Canyon, but though the gradient is comparatively gentle, it still goes up. The bright granite peaks and ridges of the Great Western Divide loom immediately before us as a reminder that whatever ease we may feel in our legs is but temporary respite. The worst is yet to come. Today, we will climb to the crest of the divide.

The uppermost Sphinx Lake is nestled deep in a rock-strewn bowl at eleven thousand feet. It is dammed and encircled by the freshly hewn stone of a horseshoe-shaped glacial moraine. A Pleistocene glacier would have left its debris heap way down at the mouth of Kings Canyon, not up here in the alpine zone. The boulders of such a moraine would be rounded from being tumbled along on their journey in the glacier. But the debris here is still angular. This moraine was left by a small, recent glacier, probably during the Little Ice Age. Since around 1750, the ice has been in retreat. Today, only a small, shrunken permanent snowfield glints in the sun above the moraine, mostly buried in debris.

Above the lake, the Sphinx Creek watershed ends in a steep talus-covered headwall, at the top of which is an obscure pass that we call Sphinx Pass. We split into two groups, don our helmets, and set out, one group at a time, ice axes in hand, up the clinking stones.

I lead the first group up the pass. There are four of us in total. We move across the unsettled talus with surprising delicacy and efficiency, given the loads we bear. If we didn't know that there was yet another pass beyond this one, I am sure our pace would slacken. But knowing we still have a ways to go before camp, we clench our teeth and persevere. In places, the talus gives way to open snowfields, where the going is easier. Here, we kick steps into the soft surface snow and find good purchase in the firmer snow beneath. Slowly the top of the pass draws nearer. After fifteen minutes, I give the signal for the second group of four to start. My trusted field assistant Will Howard, affectionately known to our group as "Granny," leads them up.

At 12,100 feet, we reach the pass and thankfully cast down our packs. Everyone is slack-jawed at the view into the interior. Gone are the canyons and forested slopes of the montane forests. Before us is a world of stone and snow, an impossibly endless jumble of high mountain peaks, ridges, arêtes, horns, granite basins, and frozen lakes. About a mile away, we can see another pass right at eye level, forming a conspicuous col in the southwest ridge of the cathedrallike Mount Brewer. Between us and the pass is a basin that drops several hundred feet before spreading to a wide, irregular floor and rising again steeply toward Brewer. To get to the pass, and our camp, we must descend, traverse the basin, and make a final ascent. For the first time since our arrival in the Sierra Nevada, we feel what it means to be in truly remote, rugged, deep, wild mountain country. The price for such remoteness is collected in the form of pain and sweat. We shoulder our packs and make for the pass.

We descend quickly over granite benches and soon find ourselves making our way across the basin floor. It is far from flat, as it seemed from above, and we carefully pick our route, traversing along the upper edge of the basin bowl, careful not to get drawn into drainages or seduced into following the downward trend of the landscape. We know well that whatever we descend we must ascend before the day is done. Once below the southwest ridge of Mount Brewer, we find the col that is our pass and work our way up a series of snowfields and steep granite ledges toward the opening. The instant we begin ascending we slow considerably. Legs turn to sandbags and lungs seem to collapse. We move like slugs upslope, slowly, steadily, and with determination. As we approach what seems to be the top of the pass, I inform the group that there is a false top, on which two lakes sit, and the true top is past the rocks and snowbanks above the lakes. They take it well, glad enough to hear that camp is near. We arrive and find the lakes partially frozen, traverse to the narrow isthmus between them, and steadily rise up and over the ledges and snowbanks on the far side. Once on top, for the second time in a day a whole new world of mountains opens up before us.

We gather on a wide, flat slab of granite, drop our packs, and circle up. We know we have done something extraordinary on this day, having climbed so high and come so far over such rugged terrain. Amid weary laughs and congratulations we drop our packs. Spontaneously, we all raise our ice ax shafts, join their points together between us, and give up a jubilant hoot to the sky. Then, one by one, we each wander off on our own to take in the place.

From where I sit, the world plunges and heaves into a hundred gray mountains splintered and smashed by age and ice and vigor. Crashing valleys are cut by overflowing cold, white, rushing creeks that bring the ice of white snowfields hung up like tapestries to the parched lowlands that are now and finally ranges and ranges away. Here, there is no man-made thing in sight, nor evidence of a past human presence of any kind. There are no fields or old hunting grounds or even the slightest artifact such as an obsidian chip. There are no signs of verdant life or the faintest suggestion of its ease. In my whole field of vision, there is not even a single tree. I sit through the changing light of evening and watch as the almost-storms of afternoon become torn and tattered and illuminated into a thousand hues of peach, orange, rose, and purple by the lowering sun. Only with the setting of the sun do I recognize the faintest sliver of the new waxing crescent set low in the western sky. Night falls over the wild heart of the High Sierra.

The Sierra Nevada forms a unique link in the greater chain of the North American Cordillera. It is conspicuously and uniformly granitic, uplifted as a contiguous block of intrusive igneous crust. It bears evidence of extensive Pleistocene glaciation yet today hosts but a scattering of dwindling glacierettes that are little more than permanent snowfields. The entire range basks in the Mediterranean climate of greater California, and while in winter the range is open to the full brunt of Pacific storms, and snowfall has been known to exceed seventy feet in a year, summers bring notoriously cloudless deep blue skies, and dry air that lasts for months on end. All told, over 95 percent of the precipitation falls between October and May, and over 90 percent of all the precipitation falls as snow. It is the southernmost extension of alpine and boreal habitat that has discernible links to the North. It is a sweeping island of wilderness sixty miles wide and hundreds of miles long, set in the hive of the thirty-five million people of California. It was the combination of vast expanses of granite, well-defined glacial lakes and landforms, long-lingering snowfields, and predictably brilliant summer weather that led John Muir to dub the Sierra Nevada "Range of Light." I have seen these mountains live up to that name a thousand times. Today is no exception.

The Sierra Nevada is over four hundred miles long, extending from just north of the Feather River country, at around forty degrees north latitude, to the Garlock Fault, at around thirty-five degrees north. What is particularly remarkable about the physical geography of the range is that it is all one contiguous geologic unit: a massive, single-fault block of granitic rock tilted gently toward the west. Averaging sixty miles in width, this fault block forms a long spine along eastern California, paralleling the northwest-southeast trend of the coastline. It is by far the longest contiguous mountain range in the lower forty-eight. The Tetons, for example, are less than 50 miles long. The Front Range of the Colorado Rockies is around 150 miles long. The Sierra Nevada, for comparison, is nearly three times as long. The combined ranges of the Rockies, as well as the combined ranges of the Cascades are more extensive, but neither of these mountain systems are contiguous.

The range tilts to the west at a relatively low gradient, averaging two to four degrees. From the perspective of California's Central Valley, which binds the range to the west, on a clear day the west slope of the Sierra appears as a great green wall crowned with snow-tipped, barren granite peaks. On the ground, however, the slope is not a wall at all but rather a forest-clad ramp fifty miles wide. The slope rises from the flat plains of the Central Valley in a series of undulating, oak-spattered hills, and ascends gradually through a series of lower, middle, and upper montane forests to the open high country above. Water on this side of the range flows down the long west slope to the Central Valley, which drains into the Pacific via San Francisco Bay.

To the east, the range drops in an abrupt escarpment to the interior deserts. Here the gradient is steep, averaging over twenty-five degrees as the mountains fall in convoluted ridges and valleys to the flat lands below. Beyond the desert basins that bind the range to the east are more mountains that parallel the Sierra Nevada. Each of these ranges—and there are scores of them—is separated from the others by wide basins, such that on a clear day the ranges extend in successive horizontal lines to the horizon. Here is the Basin and Range, bounded by the Sierra Nevada to the west, and the Rocky Mountains to the east. Water that falls on this side of the range drains into a series of terminal, saline lakes in the desert basins. From these lakes water evaporates back into the atmosphere and never reaches the sea.

The Sierra Nevada attains its greatest height in the southernmost part of the range. Here, the main crest is dissected by the defile of the Kern River, and the result is two parallel crests trending northwest to southeast. To the east is the main crest, culminating in Mount Whitney at 14,495 feet. South of Whitney, the crest dives into the trees, to rise above them again only at the alpine island of Olancha Peak, the southernmost extension of alpine habitat in the North American Cordillera. To the west, the main crest is nearly matched in height by the Great Western Divide, which extends from the vicinity of Mount Brewer in the north to Florence Peak in the south before it, too, bows below the trees.

To the north, the crest of the range is united at the Kings-Kern

Divide and extends northward as far as the eye can see and well beyond. The crest stays high through the Palisade and Evolution groups, with a dozen peaks exceeding fourteen thousand feet and scores more between thirteen and fourteen thousand feet. The crest's only significant gap occurs near Mammoth Mountain, and a discernible main crest is temporarily lost amid the black metavolcanic rocks of the Ritter Range and the buff recent volcanics of the San Joaquin Ridge. North of Donahue Pass, the southern boundary of Yosemite National Park, the crest picks up again and continues north over rounded metamorphic summits, paralleled to the west by the serrated granite peaks of the Cathedral Range, and the multicolored ridge of the Clark Range. Beyond Yosemite, the sierran granite is buried beneath tertiary volcanic rocks, though the alpine crest continues unbroken to Sonora Pass. Beyond Sonora, the High Sierra becomes lost in the trees, periodically raising its head above the foliage all the way to Lake Tahoe. Beyond Tahoe, the range is wooded and extends over a hundred more miles to the Feather River country, where the younger volcanic rocks of the Cascades bury and obscure the Sierra Nevada for good.

The long, contiguous spine of the Sierra Nevada forms a natural north-south highway along which montane and alpine plants and animals can move freely. Over millennia, the extent of montane and alpine habitat has continually fluctuated with the cycles of climate change and glaciation, but throughout these changes the corridor has always been there. Over the years, countless plants and animals, some of which are long gone from the Sierra Nevada, others of which are still well established, have used it. It is the subtropical Sierra Nevada's link to the great expanses of the arctic and boreal north. It is why mountain sorrel, *sibbaldia*, and wolverines, all of which have their origins in the North, occur at thirty-five degrees latitude.

While the Sierra Nevada serves as a migration highway for plants and animals with origins in the North, from another perspective, together with the Cascades, it forms one of the most significant natural boundaries on the North American continent. It is a hydrological, a climatological, a biogeographic, and a cultural divide.

When moisture-laden winter storms waft in from the Pacific, the airmasses cool as they rise up the western ramp of the Sierra. Cooling leads to condensation of moisture vapor, which eventually falls as rain or snow on the west slope. Often these airmasses will pool on the windward side of the Sierra, storming for days before they are finally pushed up and over the crest of the range. When finally they do descend the east side, they warm rather than cool and collect moisture vapor from the surrounding landscape rather than condense it out as precipitation. The result is a pronounced rain shadow on the east side of the range, where deserts splay out for hundreds of miles.

A diverse climate leads to diverse natural communities. The broad west slope is characterized by luxurious montane forests of pine, oak, fir, and hemlock. The flora is decidedly Californian, characterized by a high percentage of endemic species and by diversity among plants high enough for species to number in the thousands. Trout swim the waters, deer browse the hillsides, and black bears bumble along through the woods. To the east, the montane forest is compressed into a slim elevational belt, with many of the species present in the west missing entirely. The forest is drier, more open, and less productive, and below seven or eight thousand feet, it gives way to desert flora common in the Great Basin.

Indigenous human cultures east and west of the Sierra Nevada developed in isolation and became as different as the natural communities they lived in. To the east, the Washo and Piute-Shoshone tribes had relatively small populations, were mobile, and subsisted largely on pine nuts, fish, small game, and alkali fly larvae. To the west, where resources were more abundant, the Maidu, Miwok, and Tubatulabl, among others, were more numerous and generally more sedentary, subsisting primarily on fish, game, and the variety of wild plant foods available to them. The high passes of the Sierra Nevada served as natural trade routes. The peoples of the east side traded pine nuts, sinew-backed bows, paint, pumice, and obsidian, while those of the west brought skins, baskets, berries, acorns, and shell beads. The high country also served as a refuge from the heat of the lowlands. Native communities on both sides of the mountains knew the high country well and had established summer camps throughout the range.

Euro-Americans were faced with the same mountain barrier as indigenes. For years, the Sierra Nevada effectively cut California off from the larger continent. The fur trade, immigration, and later the California gold rush eventually provided incentive for exploration of the range, though this was largely limited to negotiating passage up and over the mountains. Jedediah Smith and a group of fur trappers crossed the range at Ebbet's Pass in 1826. Joseph Walker made several well-documented crossings, both at Walker Pass in the southern part of the range and through what is today the Tioga Road corridor in Yosemite. Exploration of the range itself began in 1863 when the California Geological Survey, under the direction of Josiah Dwight Whitney, began its systematic study of the High Sierra. Throughout the eleven years before the survey was brought to its official end, the team included such well-known individuals as William Brewer, Charles Hoffman, Clarence King, James Gardner, and Richard Cotter. All told, the surveyors covered a lot of ground, studying and mapping the High Sierra as they went.

John Muir's name has become more synonymous with the Sierra Nevada than any other. Muir was a Scottish immigrant from Wisconsin. He first came to the Sierra in 1868 as a shepherd, using this temporary occupation as a means to get himself into the mountains. In subsequent years, Muir did what he could to earn the daily bread that enabled him to continue exploring the mountains he came to love. He was an outstanding observer, and his observations led him to perceive unity and interrelatedness among all aspects of the land. Most noted of these are Muir's observations and studies of glaciers in the range, and the all-important role they played in the shaping of the Sierran landscape. Most of Muir's observations, however, were on a smaller scale: the falling leaf, the banner clouds, the water ouzel. Muir was a visionary, a sage, some would say a mystic.

The legendary Norman Clyde who spent more time in the High Sierra than any before him. From 1920 to 1946, Clyde made well over a thousand summits, over a hundred of which were first ascents. Unlike other mountaineers, however, Clyde was not just a visitor. Though he held a bed in Big Pine, in the Owens Valley, he lived in the mountains, making the Sierra Nevada his home and living long to reflect their awesome nature. Like most of the small number of men of this sort, Clyde was quirky. He read the classics in Greek by the fire, regularly carried hundred-pound packs, had inexhaustible physical stamina, and was known to be "a proud and sensitive man, unable to grasp modern thinking." He was said to be the last of his kind, a dying breed.

Human activity has undoubtedly increased in the Sierra Nevada, but as in the days of old it remains seasonal, and with the notable exception of the highly developed Tahoe region, there are still no permanent human settlements in the range. Today, nearly every acre of the alpine zone, and much of the montane forest around it, is protected as either national park land or federally designated wilderness. And while the population of the state of California swells to a greater number than most countries, the Range of Light remains a stronghold for the wild.

Light fades and night cloaks the mountains. But even on this night, moonless once the tiny silver sliver has set, starlight is reflected off the pale granite and lingering snowfields, and the Sierra Nevada seems to glow. Camp is silent, the others all burrowed in for the night. I quietly unpack my sleeping back, open my bedroll over a clean granite slab, and fall asleep under the stars.

Ascending Brewer Pass

# III

There is more looming stone; marbled, thunderous skies; metallic sheen of lakes; and moon-colored snowfields in this place than anywhere else I have been in the Sierra Nevada. Beneath unusually cloudy skies of afternoon, every conceivable shade of gray is present here, given greater depth by the pure black of water streaks and the white of snow, all of it washed in a pale light, like an old sepia-toned print. But though the shifting, soft, gray light is beautiful, I wish the clouds would dissipate. I wish the weather would go away. Just for a day.

We have come over Longley Pass and now make camp on the shore of Upper Lake Reflection. Before us looms the great granite wall of the Kings-Kern Divide, the highest and most rugged watershed divide in all of the range. Atop the divide, the splintered pyramidal form of Thunder Mountain rises to over 13,500 feet, the highest mountain for many miles in all directions, one of the most remote summits in all of the Sierra Nevada, and a kind of centerpiece of this rugged granite landscape, a focal point for our exploration of the range. Tomorrow we hope for clear weather, to somehow awake to the azure summer skies the Sierra is famous for, rather than the marbled gray that has overcome the mountains every afternoon for the last week. Tomorrow we will make for Thunder Mountain. But as the days of threatening afternoon thunderstorms have passed, I have thought less and less about the possibility of completing the climb, and more and more about the learning experiences that will happen if we give it our best and retreat when the weather says it is time to do so.

I sleep restlessly.

In the morning I am struck by an uncanny feeling that the weather will hold. We breakfast and pack up, but linger around camp awhile, waiting for the snow surface to soften up a bit before venturing out onto it. We did not bring crampons on this trip and thus far have not regretted that decision. Today, however, with a potentially limited window of time in which to ascend and descend the mountain, waiting for the snow to soften has me anxious. The wind is whipping and the air is cold. I am thinking about those crampons twenty miles away, in a duffel in our van, and wishing they were here.

We set out at 8:25 A.M. and skip across granite slabs and boulders to the foot of the first riser of the climb. The snow is bulletproof, but we can't wait any longer and will have to cut steps with our adzes. Slowly we ascend, the leader cutting carefully placed, large steps and the rest of us following, aiming for a ramp of granite that is free of snow, hoping it will allow us swift travel up the riser. As we approach the rock, I relax. It extends upward at an angle of around thirty degrees for several hundred feet. One by one, we step off the snow and head up the ramp.

We climb 750 feet and reach the top of the boulder-strewn ridge that separates our camp from the Kings-Kern Divide. From here, we look out over a hidden basin to the massif of Thunder Mountain, and preview our route. We will cross the empty glacial basin before us, ascend a thousand vertical feet of snowfield to the scree chutes and cliffs of the headwall of Thunder Col, then continue up the broken back side of the mountain to the blocks and notches of the summit ridge. Once on the summit ridge, Will and I will set fixed lines along the final traverse to the uttermost summit block. The plan is a bold one, especially for a group of eight and considering the weather pattern over the last week. But it is not at all impossible, and with efficiency and luck we might make it all the way to the summit and back. I urge my company to observe the weather but not to speak about it out loud.

We descend the ridge, cross the basin, and start up the snowfield at a brisk pace. Here, the snow has begun to soften, and as the angle increases, we are able to kick steps more easily. The students with the heftiest boots lead the way, and the going is easy. As the angle steepens still more, we plunge the shafts of our ice axes into the snow more deliberately, though the runout is good and a slide would eventually bring one down to a wide, low-angle apron of white a few hundred feet below. But the snow is coarse and not bonded together here, like a bed of ball bearings. Just when I begin to get nervous, we step onto the scree and rocks of the headwall.

We divide into two groups and traverse one group at a time, moving in a tight horizontal line with a slight incline, careful never to be on top of one another. Even the tiniest steps send sand and stones sliding downslope and onto the snow surface below. When the first group reaches the top of the headwall, they break up into pairs of two and negotiate narrow chutes through the cliff bands. Once they are all on top and we get the word, my group follows.

A brisk wind whips across the top of the pass, where we gather to eat and drink and prepare our climbing equipment. The mountain rises in a heap of broken stone from here, and the remainder of the ascent is 80 percent steep, loose talus slog and 20 percent technical climbing on high-quality sierran granite. I figure we will make short work of the talus slog. It's moving all eight of us through the technical part on top that will take the better part of the afternoon.

We split into two groups again for the climb to the summit ridge. Granny and I lead the first group, along with Ingrid Dahl and Chris Marshall, the two students with the most climbing experience. Our plan is to get a head start to the summit ridge and start setting up fixed lines while the second group climbs up the talus.

The mountain is even more broken than I remember it, and though we carefully pick our way up, rockfall is more of an issue than I would like. About halfway up to the summit ridge, my protective instincts kick in and I decide to send the first group up to the summit ridge and wait for the second group. They come up quietly and suredly, having picked a better route than we did. After I join them the climbing gets easier and the rock more solid. We climb methodically but efficiently to the summit ridge.

Granny has an anchor ready when I arrive. I tie into the end of a rope and he gives me a belay. I traverse a series of steep blocks and descend into a deep notch that separates the southern section from the northern section of the summit ridge. The notch, reached by some long vertical downclimbing moves, is choked with snow, and I have to stamp a ledge into the snow to step down onto. Once there,

I traverse over to the base of the northern section of the summit ridge. Here, I build an anchor, secure the rope, and wait for Granny.

Granny follows, sliding a prussic that is tied into his harness along the now fixed line as he moves along. His tall form allows him to make the downclimbing moves into the notch with a grace and ease that I could never attain. When he arrives, he clips into the second anchor and sets up to give me a belay on the second line. I tie into the end of it, unclip from the anchor, and begin part two of the ridge traverse. As I move across the ridge, I cannot believe that I climbed this same route solo just two years before. I would never do that now. I've lost my nerve for it. I don't want to die in the mountains.

The summit block is as I remembered it, except that this time I am climbing with a rope. I thankfully insert two camming devices into a prominent crack and clip the rope to them. Then I make the bold moves to the top. On that wide, warm block of stone, I build a third anchor and fix the line. Finally, I tie into the anchor, leaving about ten feet of slack, just enough to give me the freedom to make the final move to the very tip-top. It is a space barely big enough for a few tightly packed bodies, a precarious place of lofty granite blocks that drop into oblivion. I haul myself up there. Beneath an angular chunk of granite is an old summit register. In it I discover the script of Norman Clyde, who made at least five ascents of Thunder Mountain between 1928 and 1936. Jules Eichorn's signature also graces the pages numerous times, along with other icons from the history of mountaineering in the Sierra Nevada. I humbly scrawl my name onto a page, take a few breaths, and dangle myself back down to the main block.

As the others traverse and ascend to the summit block, we cluster together in the sun and bask in a pervasive, well-earned sense of accomplishment and beauty on this notoriously unforgiving mountain. One by one, the rest of the group clambers up to the topmost summit block, each of us on a little solo pilgrimage to such a distinguished spot. Thunder Mountain is not a mountain to be taken lightly, and certainly nothing to be conquered. Rather, the mountain allows one to climb it, and we do so submissively. Today, we approached the throne of God humbly and so were able to visit for a while.

In all directions the view is of a vast granite mountain landscape that tumbles away to the horizon. In fact, I am hard pressed to find anything in the scene that is not granitic. The southern Sierra is a world of granite, shaped and sculpted by glacial ice into peaks and horns; knife-edged arêtes and saddle-like cols; flat basin treads filled with sparkling blue and turquoise lakes; steep, slabby risers laced with white, frothing waterfalls; heaped alpine moraines; and deep, steep-walled U-shaped glacial valleys. Here is a piece of one of the largest and most significant bodies of granite on the planet, and one of the most perfectly hewn postglacial landscapes to be seen in the world. Amazingly, while all this granite was once the basement of an immense chain of volcanoes, scarcely a scrap of those ancient volcanoes is visible from our perspective. And while glaciers once covered the range from end to end and sculpted the range to its present form, not a single glacier can be seen from here.

The story of the Sierra Nevada follows the story of the entire American Cordillera, and reaches back in time to some five hundred million years ago, when all of the continents on earth were converging to form the great supercontinent known to us as Pangaea. At this time, California and much of the western portion of the Americas did not yet exist. The west coast of North America ran through what is now central Utah and part of Arizona. North America moved east, toward Africa and Europe, and all along the eastern seaboard, mountains were being thrown up as the continent bulldozed its way over the existing ocean basin. In the west, however, things were quiet. Much like the eastern seaboard of both North and South America today, rivers washed sediments from the land out to sea, and a broad shelf of sediments gradually extended the continents westward. Today, much of these old Paleozoic seafloor sediments are exposed in the Grand Canyon and adjacent regions of the Colorado Plateau.

This passive accumulation of sediments along the western margins of the Americas continued throughout the assembly of Pangaea.

Meanwhile, on the eastern margin of the North American continent, Alpine-Himalayan-style mountains were being thrown up all over the place as the continent collided head-on with Africa and Europe. Pangaea, as is the case with any supercontinent, was not built to last. Fully assembled by 250 to 300 million years ago, by 215 million years ago it began breaking up. The amassing of all of the continental crust into a single supercontinent overinsulated the upwelling convection cells of the mantle, much like a lid on a pot of boiling water. Eventually, the pressure built enough to pop the lid off. Pangaea rifted as quickly as it had joined. The overinsulated convection cells split Pangaea apart in several places. One of these places was along the seam where the Americas met Africa and Europe. Here, where the continental crust split and rifts widened, seawater eventually rushed in, and the Atlantic was born. North America, freed from its brief embrace with Africa and Europe, did an about-face and began to move westward.

The westward movement of the Americas that followed the breakup of Pangaea brought an end to millions of years of passivity along the western margin of the American continents. As the opening of the Atlantic pushed the continent westward, the Americas bulldozed their way over the adjacent ocean basins and devoured them. In the case of North America, the oceanic plate next door was the Farallon Plate. The Farallon Plate, in direct opposition to the North American Plate, was moving *eastward*. As the Farallon Plate subducted beneath the overriding continent, it dove into the mantle, scraping off the underside of the continent along the way, and melted. The accumulating molten bodies of magma that resulted from this subduction soon began to push back up through the crust of the overriding continent, just inland from the actual subduction zone. These bodies of magma, called plutons, rose as teardrop-shaped distillations, melting even more of the crust as they intruded into it and thus becoming still more buoyant. Eventually, many of these plutons reached the surface of the continent, and extruded as an extensive arc of volcanoes that extended from present-day Mexico all the way to present-day Alaska. These pre-Sierran volcanoes built up with successive eruptions, and by a hundred million years ago had grown to exceed fifteen thousand feet in elevation, comparable to the Andes of South America today. The Andes themselves were just beginning to form at this same time, due to a subduction dynamic similar to that occurring in the North.

Beneath the pre-Sierran volcanoes, plutons continued to intrude into the continental crust, forming a vast reservoir of magma beneath the volcanic arc. Eventually, much of this magma cooled beneath the surface, transforming itself into an immense basement complex of granite. The formation of this body of granite was one of the most substantial granite-forming events in the history of the Earth, and outcrops of this basement occur from southern Baja California far into the Pacific Northwest and intermountain states. Most significant, this granite is exposed along the length of the Sierra Nevada.

In time, the spreading of the Atlantic sped up, and seventy million years ago it began pushing the Americas west at a faster pace. The Farallon Plate, rather than subducting, diving into the mantle, melting, and redistilling into the continental crust as plutons, began to slip beneath the continent at an increasingly low angle. Instead of melting, the oceanic plate is believed to have slid directly under the continent. As it did so, the old, relatively elastic sedimentary strata of what is today the Colorado Plateau upbowed as a unit, and the plateau was born. Further to the east, where the crust was more brittle, the underriding oceanic crust rumpled the stuff of the continent, and the Rockies rose.

While the mountain-building action moved east to the Rocky Mountain region, the warm, wet climate regime of the Cretaceous Period made short work of washing away the pre-Sierran volcanoes. The sediments from this pronounced erosional phase eventually filled in the basins on either side of the range, accumulating sediments in California's Central Valley up to forty thousand feet deep. Within a few tens of millions of years, the volcanoes were gone, and a series of low-lying, rounded hills were all that remained. But at the crests of these hills, broad expanses of granite were exposed, suggestions of the great body of granite beneath that was lying in wait.

While the Farallon Plate was subducting beneath the continent it was delivering small oceanic islands and microcontinents to the leading edge of North America and pasting them on to the continent. These accreted terranes grew the continent westward, creating much of California, Oregon, Washington, and British Columbia. Similar accretions eventually formed much of Alaska, parts of Mexico and Central America, and the western slice of South America.

By thirty million years ago, the Farallon Plate off the coast of California was completely subducted, devoured, gone. Remnants of it still remained to the north, in the vicinity of what is now the Pacific Northwest, and south, near present-day Mexico. The next oceanic plate in line for collision with North America was the enormous Pacific Plate. The Pacific Plate, however, had a different agenda. Instead of moving east, like the Farallon Plate, it was moving *northwest*. This movement was not at all in direct opposition to the movement of North America, but rather similar to it, in an oblique sort of way. When the Farallon Plate was devoured and North America butted up against the Pacific Plate, it did not subduct, but rather grabbed on to North America and began to carry it northwest.

In the Pacific Northwest and Mexico, where the remnants of the Farallon Plate still existed, subduction resumed, and it continues to this day in the regions of the Cascades and Mexican volcanoes. In California, however, where the Pacific had grabbed hold of North America, the tectonic regime changed completely. Instead of subducting, the Pacific Plate began dragging the continent northwest, stretching and eventually ripping it apart in the process. Like a gigantic piece of taffy, the crust stretched and thinned to accommodate extension, its width increasing over a hundred fifty miles in some places. Eventually, however, the strain was too great, and the crust began to break. The main suture of the Pacific and North American plates broke into the San Andreas Fault system. To the east, scores of lesser faults eventually formed, each of these paralleling the San Andreas to the west. The crust broke into a series of long blocks whose edges were bounded by the newly formed faults. Some of these blocks tilted upward to form mountain ranges. Other blocks sank as the crust continued to spread. At the edges, along the faults, molten material from the mantle seeped and exploded outward. The Basin and Range Province was born.

The granitic basement of the pre-Sierran volcanoes held together when the crust broke, especially in the region that is now the Sierra Nevada. Major faults formed along the eastern flank of the partially exposed basement, and some along its western edge as well. This immense block began to uplift and tilt, first in the south, and eventually in the north, ripping apart cleanly along its eastern edge to form the eastern escarpment of today. To the west, however, sediments from the long-eroded pre-Sierran volcanoes anchored the block down, so it rose like a trap door, open to the east. Because the basement complex, for so long the root of the pre-Sierran volcanoes, was so deep, its rate and amount of uplift exceeded any of the other uplifted blocks of the Basin and Range. And thus the modern Sierra Nevada was born.

The extension of North America does much to explain the overall mechanism of uplift for the Sierra Nevada and how it differs from that of any other range in the Cordillera, but it does little to explain how the range was excavated into its current form. When the Sierra rose, it did so as a massive granitic plateau. Carved only by water, the profile of the range would have looked broad and gently undulating, much like the unglaciated White-Inyo Range of today, just across the Owens Valley from the Sierra Nevada.

Pleistocene ice changed the character of the Sierra Nevada forever. At the height of the glacial periods, a mountain icecap in excess of 275 miles in length and 4,000 feet thick blanketed the range. From this central dome of ice, a myriad of lesser valley glaciers extended like tentacles to the west and the east, terminating at the mountain fronts. For over three million years, the glaciers periodically plucked, scraped, scoured, and carried away thousands of cubic kilometers of material, working in conjuction with rockfalls and slides to patiently sculpt the landscape of today's Sierra Nevada. The topography of the range was dramatically changed from a high mountain plateau to a jagged expanse of high, vertical peaks and deep, troughlike valleys. The tilt of the range was also exaggerated, as the larger western valley glaciers deposited huge amounts of debris in the foothills and weighted the block down even more. In the east, however, the range continues to rise to this day. In 1872, a single earthquake caused over twenty feet of lateral movement and thirteen feet of vertical movement in the southern Sierra Nevada. And though the glaciers of the Pleistocene are gone, they will be return.

Glaciers. Granite. The evidence on one imprinted in the other surrounds us on all sides. Our attention is returned to our situation by the ominous clouds building to the northwest. "Looks like the Ice Age is back," Granny jests. The storm will be here in a couple of hours. We have time, but not that much time.

One by one, we retrace our steps across the traverse. I sweep up, and pull out anchors and protective gear along the way. We move with fluidity. When the first group of four reaches the first anchor, they set out down the broken back of the mountain for the col. I make my way back across the ridge and descend into the notch. At its bottom, the snow ledge I stamped out two hours ago is nearly wasted. I step onto it hesitantly. Just as I step off the snow and back onto rock, the platform breaks and crumbles away. I hoot out loud and begin the series of long moves that lead out of the notch.

At the col, Granny, Ingrid, Chris, and I join the others. Everyone seems lit up, ecstatic, as if infiltrated by the spirit of this place. We enjoy a few moments of bliss before heading back down the scree and snow of the headwall.

We are relieved to reach the upper slopes of snow. It has softened considerably since our ascent. After a few words of caution to each other, we glissade and plunge-step down the entire way, like a bunch of starry-eyed children on the loose. For a thousand feet we glide and run, clear down to the bottom of the basin. What took us hours on the way up takes us minutes on the way down.

Camp. Evening. Thunder Mountain is entirely surrounded by ominous gray clouds and an eerie yellow light. Thunder rolls from deep inside the hidden vault of the sky, and any evidence that the day was bright and blue has been banished from the world. It is strange to me that there has been no day in the last ten when we could have done what we just did, how we left camp this morning with a literal bow of reverence for the mountain, expecting perhaps to get closer to it, but never to sit upon its uppermost heights, how even until the last moment that we stood upon its summit ridge, the sky was blue, and the hour we left, it was entirely gray, how now the world shakes with thunder and the mountain is lashed with wind and rain.

I watch the tiniest mouse I have ever seen pad around the boulders of our kitchen. Its eyes are deep and black and show no sign of anything but the intensity of being alive and in the moment. The first few drops of cold rain send it running for cover. The patter of rain drops brings the scent of rain on granite, a smell so fresh it is almost astringent, but so alive it makes me breathe deeply over and over again.

I follow the mouse's example and retreat to my tent. Within minutes, it seems as if a sheet of water is crashing down all around me. I peer out, and the mountain is totally gone, the last trace of its gothic form enveloped in an esoteric, impenetrable wreath of thick, dark clouds. The storm pushes downdrafts across camp, wicked pumping winds that send our stoves clattering across the granite. I wonder about my new friend the mouse, and guess that he of all creatures is okay. Soon, the sound of water beating down on the tent is all there is, all I can think about. And somewhere out there, in the tempest, swarmed by laughing dragons, is the most mystical mountain in the Sierra Nevada.

Thunder Mountain Ridge

# IV

It is late June, but here in the Sierran alpine zone it is just spring. The melting snowbanks are still thick, and if not for the meltwater gurgling beneath them and the brilliant patches of spring-green sedges and herbs between them, it would be easy to think it was still late winter. But closer inspection reveals an explosion of life from the emerging meadows, rocky edges, and still-soggy gravel flats. Insects buzz conspicuously over the landscape, busy in the sunlit day. The sweet smell of nectar is on the wind. The alpine bloom is under way.

We step gingerly across the landscape, trying to stay on rock as much as possible. In the High Sierra this is easier than in most places. The landscape is over 60 percent exposed rock. Today, it is still thirty percent snow. But in between are alpine gardens of tender plants just awakening to the warm sun of the growing season. We have been observing them closely, noting the composition of the alpine flora, watching the progress of the alpine bloom relative to that of the montane forest below. Down below treeline, the bloom is in full swing, the whole ecosystem afire with the energy of summer. Here, above the trees, that fire is just being kindled, soon to burst into flame.

When the flowers of the tundra bloom, they do so suddenly, almost with reckless abandon. Because the growing season is so short, there is little time for staggering the bloom over many weeks or months, for after the bloom there must be enough weeks left in summer for fruits to mature, and seeds to ripen and be dispersed. Species tend to have synchronous blooms, meaning all of the individuals in a given population burst into flower at the same time, often even the same day. The effect can be mesmerizing, as you find yourself sitting in a field of thousands of bog bilberry flowers one week, and surrounded by thousands of alpine shooting stars the next. After a few summers up on the tundra, however, the bloom cycle gets more predictable. Saxifrages tend to be among the earliest bloomers, sometimes even starting to open their flowers beneath the snowpack. Heaths such as bog bilberry and bog kalmia can be early too, though seldom as early as some of the saxifrages. Penstemons, shooting stars, wild buckwheats, and phlox are conspicuous in early July, and represent the peak of the bloom. Gentians and many composites bloom through August and into September, taking advantage of a time when most other plants have already gone to seed. There are hundreds of others; these are just a few.

Here, things are happening fast. On the gravel flats, pussypaws, alpine everlasting, and wild buckwheats, long smashed by snow and winter winds, are beginning to show shades of pale, silvery green, evidence that the energy held in their root systems all winter is moving back up into the aboveground parts of the plant. The telltale leaves of penstemons show only the earliest signs of flower buds. Out on the meadows, the green leaves and tiny pink urns of bog bilberry are out, taking advantage of early season moisture that is sure to be gone by midsummer. Nearby, a woody mat of arctic willow sends up its first green catkins. In wetter areas, old, brown sedges are still pressed to the ground, freed of snow just days ago. But even in their midst, the spring-green leaves of perennial herbaceous wildflowers poke through. In sheltered areas, scrawny, shredded whitebark pines, their needle tips orange from winter burn, push the absolute limit of treeline to 11,500, even 12,000 feet.

The Sierran alpine flora is like no other. It is hailed by botanists as the largest and richest of any alpine area in North America. The number of plant species in this single mountain range rivals that of the circumpolar tundra from Alaska to Greenland to Siberia. Botanists have recognized over six hundred species of vascular plants here, over two hundred of which are unique to the alpine zone and do not mix into other communities at or below treeline. Only 14.2 percent of these species have their origins in the tundra of the circumpolar North, by far the lowest percentage of any alpine tundra region in

North America. Another 3.8 percent of these species have their origins in the northern boreal forests, again the lowest of any alpine region in the continent. Eighty-two percent of these species are uniquely alpine, with little or no connection with the north. Of these, 17 percent are found nowhere else in the world but the Sierra Nevada.

Why so many biogeographic superlatives? Common sense provides one answer. The Sierra Nevada is over two thousand miles away from the nearest Arctic tundra and over twelve hundred miles away from the nearest slice of the northern boreal forest. This is farther from the North than any alpine region of significant-size on the continent. If the skyline migration corridor of the Sierra-Cascade axis were contiguous, such distance would pose much less of a problem for plant migrations, but the corridor is not contiguous. Along the seven-hundred-mile span of the Cascades, it is frequently broken, and montane and alpine habitat are as isolated as a series of sky islands in a sea of lowlands. Between the North Cascades and the Sierra Nevada, these islands are relatively small. For each gap in the corridor, the number of plant species that makes it south gets smaller and smaller.

Another answer requires a bit more ecological meditation. The Sierra Nevada has a climate unlike any other high mountain range in the continent. In summer, with the exception of relatively rare thunderstorm cycles, climatic conditions are desertlike. In fact, if not for the meltwater of late-lying snowbanks, there would be precious little water available during the critical months of the growing season, and the Sierra Nevada *would* be a desert. Today, many species in the Sierra Nevada are either the same as or have close affinities with species of the Great Basin desert and semiarid grassland regions adjacent to the Sierra Nevada. These species with lowland origins did not have such a long journey to make to get to the High Sierra. Many of them, especially desert plants such as wild buckwheats, were preadapted for success in the alpine zone due to their existing ability to store carbohydrates in their root systems during hard times, as well as their moisture-conserving growth forms and leaf adaptations. Others, including an array of winter-blooming species (not uncommon in deserts such as the Great Basin where most of the annual precipitation falls in winter), had already evolved to undergo metabolic processes in cold temperatures, and the switch from the desert winter to the alpine summer was easy. Such preadaptation may even have led to nearby lowland species outcompeting and displacing some of those northern species that *did* make it down to the far-flung Sierra.

As we meander across the tundra, we focus on five plant families that together represent the unique biogeographic situation of the Sierra Nevada relative to the North American Cordillera and the North in general. These are *Polygonaceae*, the buckwheats, *Scrophulareaceae*, the snapgragons, *Ericaceae*, the heaths, *Salicaceae*, the willows, and *Pinaceae*, the pines, spruces, firs, Douglas firs, and hemlocks.

The wild buckwheats that grow in the Sierra Nevada, primarily of the genus *Eriogonum*, have their origins in the adjacent lowlands and are common all over the deserts of the Great Basin. As alpine plants, they are most common and widespread in the Sierra Nevada and a few scattered basin ranges. Elsewhere in the Cordillera, where moisture is not a limiting factor, the eriogonums are absent, outcompeted. But in the dry conditions of Sierran gravel flats, where snowmelt water drains off early and the thin soil is parched for most of the growing season, they thrive. They hug the ground in cushions or mats of tiny, fleshy leaves, variously coated in powder or hair. Their flowers are miniscule, lacking petals, and it is difficult for all but the most skilled botanists to tell apart the half dozen species that occur here. We are becoming skilled botanists, as wild buckwheats characterize the gravel flats, and extensive gravel flats characterize the Sierran alpine zone. Nowhere else in the Cordillera, or in the entire Northern Hemisphere as far as I have seen, are gravel flats so extensive as an alpine habitat. Within a twenty-foot radius of where we are crouched, three different species occur in the gravel flats, including oval-leaved eriogonum, cushion eriogonum, and frosty eriogonum. As we travel north, we expect to leave the gravel flats behind. As we leave the gravel flats behind, we will say goodbye to the eriogonums.

If the eriogonums epitomize the unique conditions of the Sierra Nevada alpine zone, the *Scrophulariaceae* family, also known as the snapdragons, epitomize the mid-latitudes of the Cordillera. These are among the most conspicuous and charismatic of the alpine wild-flowers, including all of the penstemons, elephant heads, louse-worts, paintbrushes, and monkey flowers. A proliferation of species in this family occurs throughout the Sierra Nevada, Basin Ranges, Rockies, and Cascades, but interestingly, their diversity decreases as you go farther north. Although present throughout the Cordillera and in the North, they seem to have come into their own in the mid-latitude mountains and diversified considerably here. In the Upper Lake Reflection basin alone, even a cursory survey of the area reveals at least eight species of snapdragons, including Sierra penstemon, Davidson's penstemon, greater ele-phant heads, lesser elephant heads, alpine paintbrush, Lemmon's paintbrush, alpine monkeyflower, and seep-spring monkeyflower. Of these, Sierra penstemon, Lemmon's paintbrush, and alpine paint-brush are unique to the Sierra Nevada, while Davidson's pen-stemon, both species of elephant heads, and both species of monkey flower occur north along the Cordillera.

The heaths of the *Ericaceae* family epitomize both the tundra and the boreal forests of the North perhaps more than any other vascular plant family. This should come as no surprise, as the heaths spe-cialize in the acidic soil conditions common to both conifer forests and boggy solid where moisture levels are high, temperatures low, and decomposition rates slow. These conditions prevail almost everywhere in the North, and, not by mere coincidence, all over the high mountains of the North American Cordillera as well. Among alpine plants, it is the heaths that most clearly illustrate the bio-geographic link that the mountains of the Cordillera have with the North. Blueberries, bilberries, cranberries, huckleberries, laurels, and a wide variety of heathers are all constituents of this family, and one could travel to just about any location in the North or any mountain range in the Northern Hemisphere, look at the ground, and find heaths. In some places, such as the Highlands of Scotland and the skerries of Norway, entire ecosystems are dominated by heaths, and are referred to by the locals as "the heath" or "the heather". In the Sierra, they rarely cover the ground to this extent but are surprisingly diverse nonetheless. At Upper Lake Reflection we find five species of heaths, including red mountain heather, white heather, Labrador tea, dwarf bilberry, and bog kalmia. Of these, the red mountain heather is unique to the Sierra Nevada, while the white heather, bilberry, Labrador tea, and bog kalmia occur throughout the North.

*Salicaceae*, the family of the willows, is another group of plants commonly associated with both the North and the mountainous areas throughout the Cordillera. They boldly employ the strategy of winter deciduous leaves in lands where the growing season is theoretically much too short to put out new leaves every year and still have time for reproduction. They deal with this in a variety of ways, from starting next year's leaf buds the autumn prior (almost every species in the family does this), to getting a head start in spring by photosynthesizing through their bark, as aspens are inclined to do. In addition to these adaptations, willows have backup strategies in the event that a given growing season is too short or stressful to allow for sexual reproduction. They can reproduce asexually through vegetative means. Willow branches that are broken off can wash downstream and resprout in sandbars miles away. Aspens send up shoots from extensive underground root systems like grass, and grow extensive colonies of clones. In the Sierra Nevada, two common species of the willow family have their origins in the North: aspen, which grows in disturbed sites below treeline, and arctic willow, which grows commonly in moist meadows on the tundra. In addition to these northern representatives, half a dozen species unique to the Sierra Nevada and half a dozen more that grow along the Cordillera occur throughout the range.

The family *Pinaceae* as a whole is a major constituent of the temperate Northern Hemisphere. Several genera, such as *Pinus*, have diverse origins and distribution, while others are more specific to the cold regions of the North. *Picea*, the spruces, are synonymous with the boreal forest, and are the primary constituents of the largest and most contiguous terrestrial biome on the earth. They are also synonymous with the upper montane forests of the mountains of the Cordillera, as they follow the migration corridors of the mountains, crossing latitudes deep into the south. They occur in every major range in the North American Cordillera *except* the Sierra Nevada. The reason for this notable exception is simple. There is not enough summer moisture here to support them, and this is the only major range in the Cordillera where this is the case. Here, in the range of eternal summer sunshine, other species in the pine family occupy the niche that spruces typically fill. The most widespread of these is the lodgepole pine, which occurs throughout the Cordillera and the western boreal forest, but occurs in greater abundance here in the Sierra Nevada perhaps because of the absence of spruce. Lodgepole pines are among the most opportunistic of pines, and do well in such a variety of conditions that they grow all the way from below five thousand feet clear to treeline in the Sierra Nevada. In the upper montane, in the northern part of the range, they share space with the more shade-tolerant mountain hemlocks, and in the southern part of the range with the sun-loving foxtail pines. Throughout the range, another species, the whitebark pine, dominates the treeline ecotone. Among these, whitebark pine is the only species that grows high enough to occur in the Upper Reflection Lake basin. The lodgepole is the only species with direct connections to the North, mountain hemlocks and whitebark pines grow elsewhere in the Cordillera, and foxtail pines are endemic to California.

The day grows late. We eat lunch in the sun, then turn to the chore of breaking camp and packing up to hike down into the montane forest. It does not take us long, as we have become well practiced by now. In an hour we assemble, packs on, ready to go.

The descent is abrupt and rugged. We slide several hundred feet down a system of loosely connected, long, narrow snowfields, careful not to step too close to their weak edges. They lead us to the narrow, talus-choked gully through which the water drains out of the lake above. We carefully negotiate the talus, listening to the roar of water just beneath the huge chunks of stone, following the invisible water down. Several hundred feet below, the gradient of the gully lessens significantly, and we find ourselves thrashing through willow thickets. The scene is Alaskan. But there are no grizzlies hidden in the bushes, and today I am glad about that.

We find our way through the willows and emerge at the inlet to Lower Lake Reflection. Having descended a thousand feet, we find ourselves in a montane landscape in the full swing of summer. Heathers, penstemons, monkeyflowers, lupines, lilies, shooting stars, and an assortment of flowers in the rose family are in full bloom. Tall lodgepole and western white pines, along with scattered red firs emit sweet, resinous odors in the warm afternoon sun. Though not seen, birds can be heard, their songs coming from every direction. Countless bees and butterflies buzz and flutter over the ground, and smaller, unnamed insects form swarming clouds. Within minutes of stopping we are literally covered with mosquitoes. Here, at 10,500 feet, we find ourselves in an ecosystem similar to the North Cascades at 5,000 feet, or the Yukon at 1,000 feet. Although many of the species are different, the scene is much the same.

We traverse the sloping granite ledges around the west side of Lake Reflection and make for its outlet. Here we pick up a small, indistinct, user-maintained trail. The trail follows the waters of the newly formed East Creek, gushing in a series of cascades from Lake Reflection to East Lake, one and a half miles and several hundred feet below. As we descend, we pass wet streamside meadows decorated with the nodding, five-spurred flowers of crimson columbine, brilliant yellow patches of seep-spring monkey flowers, and slender, subtle rein orchids. Beneath the canopy of conifers, blue-eyed Marys form delicate purple lawns. As the afternoon waxes, we spy deer

*Phyllodoce Breweri*

browsing in the old horse meadows above East Lake. When we reach
the shores of the lake, dusk is upon us, but the energy of summer all
around us has our spirits high, and we decide to hike into the darkness and make our way down to Bubbs Creek, deep in the heart of
the montane forest.

Below East Lake, we descend quickly along switchbacks, always
within earshot of East Creek, roaring somewhere through the trees.
Eventually, the gradient lessens, and we emerge into an open area.
Fire signs are everywhere. Charred logs litter the ground. A proliferation of shrubs, including chinquapin, manzanita, and ceanothus,
grow where trees once stood, interspersed with abundant flowering
lupines. Before us, across the gorge of Bubbs Creek, the giant granite
monolith of Charlotte Dome looms to a thousand feet above us, its
clean stone washed in the last pink light of the day.

Darkness falls as we approach the southern bank of Bubbs Creek.
It is swollen from the snowmelt of the high country and its waters
are white and raging. Here we make a hasty camp in a grove of
immense red firs. Tomorrow, we will cross this roaring creek. We will
leave these trees and flowers and descend to the dusty bottom of Kings

Canyon, where pines and oaks grow out of the dry earth, and eventually even these trees give way to parched grasslands where snow never falls and alpine wildflowers would blister hopelessly in the lowland sun.

I always get a sinking feeling when I leave the montane forest and descend to the lowlands. The very presence of the hot valley grasslands and desert basins often becomes an uncomfortable reminder that the montane forests and alpine country of the Sierra Nevada are indeed isolated islands in a sea of lowlands, that however long you wander in the High Sierra, if you travel long in any one direction, you are bound to run out of space and have to come down. But I find hope to the north. Soon we will leave the lowlands behind. As we go north, the islands will grow larger and larger. We are going to where the mountains and forests reach clear down to the sea.

# V

Seven hundred miles north of the Sierra Nevada, the North Cascades stand like an enormous set of waves cast in stone, dark crests encrusted with blue glacial ice, emerald-green slopes streaked with brilliant snowfields and dark rows of conifers, tatters of Pacific clouds clinging to the landscape like well-worn shrouds. The mountains form a cluster of high, jagged, dark gray peaks that seem to rip into the clouds like claws, splitting them open time and time again, spilling out endless storms of rain and snow, feeding cracked, groaning glaciers, watering lush, flowering meadow plants growing along old avalanche chutes, feeding tall trees dressed in pale green hanging lichens. It is a world apart from the bright, open granite landscape of the Sierra Nevada to the south. It is the first strong suggestion of the mountain landscapes to come in the North.

Here, the connection to the North becomes direct, and plants and animals have been free to move around along a contiguous corridor for millions of years. Here, the great Cordilleran Ice Sheet of the

Pleistocene, extending north along the mountainous edge of the continent even to the Aleutians, came to its southern end. Here, the tens of millions of contiguous acres of relatively wild land of the northern Cordillera also find their southern terminus. From the perspectives of northern flora and fauna, glaciers, and wildness, the North Cascades are the first sure suggestions of what is to come on the journey north, as well as a decisive reminder of what is being left behind. Gone are the open, undulating granite plateaus of the High Sierra. Gone are the clean-cut, empty glacial bowls and valleys. Gone are predictably clear summer days. Gone are the warm pines, manzanita, and wild buckwheats. Gone is the ever-present hum of surrounding California. Here, the mountains are noticeably steeper and more rugged than the mountains to the south, the local relief greater that anywhere else in the lower forty-eight. The hollows of the terrain are caked with glacial ice. The sky brings grumbling weather that reaches far into summer. The vegetation on the lower flanks of the mountains is a thick, nearly impenetrable temperate rainforest. And though the nearby urban centers of the Puget Sound region buzz with human activity, it is the last significant outpost of large-scale modern western culture along the Cordillera. Here is the middle piece of our progression north.

The Cascades as a whole span some seven hundred miles from the Lassen region of California to the Fraser River country of British Columbia, north of which the range blends into the Coast Ranges of Canada. Perhaps because the Cascades are such a long and diverse range, geographers divide it up into three sections. The Southern Cascades extend from the southernmost extent of Cascadian volcanism, at around forty degrees north latitude, to southern Oregon, somewhere between forty-two and forty-three degrees north. This part of the range is most susceptible to the influence of California's Mediterranean climate, and is known for its relatively low precipitation rates, mild temperatures, and few glaciers. Widely scattered stratovolcanoes, such as Mount Shasta and Mount Lassen, characterize the topography. The enormous form of Shasta, rising conspicuously like the Japanese Fujiyama above the open expanse of lowlands, dominates the Southern Cascades, and at 14,162 feet it is the second-highest peak in the range.

The Southern Cascades blend imperceptibly into the Central Cascades, which extend from the upper reaches of the Willamette Valley in southern Oregon to the Yakima River in Washington. Here the Cascades attain their greatest height, and Mount Rainier rears its glacier-covered dome to 14,410 feet. Mount Adams, Mount Hood, and the Three Sisters also characterize the region. In the Central Cascades, the Mediterranean influence of California is markedly less pronounced, precipitation levels are generally higher, temperatures lower, and glaciers better developed.

The Northern Cascades begin at the Yakima River Valley and extend north to the Fraser. Far from the Mediterranean influence of California, this part of the range receives the most precipitation, has the coldest temperatures, and the most extensive glaciers, accounting for 75 percent of all the glaciers in the lower forty-eight. The relatively low relief of the Cascade volcanoes to the south mixes with the relatively high relief of mountains formed from continental collisions. Volcanic peaks such as Mount Baker and Glacier Peak are set among a myriad metamorphic peaks, including Mount Shuksan, Mount Challenger, the Ricket Range, Eldorado Peak, Forbidden Peak, and Dome Peak, to name but a few. Although the mountains here are the lowest among the three geographic regions of the Cascades, they have the most rugged terrain, the worst weather, the biggest glaciers, and are the most inaccessible.

Climate defines this place. The combination of subpolar latitudes, maritime position, and high elevation results in impressive, sometimes oppressive, precipitation loads. Summer offers some respite, as even the Northern Cascades are at least somewhat subject to the Mediterranean influence of California. Between six and twenty times more precipitation falls in winter than in summer. Most of this falls as snow, making the North Cascades the snowiest place in the lower forty-eight. As is the case with the Sierra Nevada, the mountains form a significant climatological barrier and cast a considerable rain shadow to the east. While the west slope at Cascade Pass is drenched in 160 inches of precipitation per year, Mazama, just twenty miles to the east, receives only 15 to 20 inches.

The climatological variation between the west and east slopes of the North Cascades has predictable influence on the natural communities of the region. The lower elevations of the west slope are cloaked in the dense temperate rainforests that the Pacific Northwest is famous for. Here, Douglas firs, silver firs, western red cedar, and western hemlock, often thickly draped in hanging lichens, form a supercanopy of up to three hundred feet in height. Beneath these giants of the forest grows a midstory of Pacific yew, big-leaf maple, vine maple, and alder. Even lower to the ground, a rich understory of ferns, shrubs, and herbacious wildflowers decorates the forest floor. Dim, green light permeates the place. Further upslope, snow falls more often than rain, and the rainforest gives way to a boreal-type montane forest characterized by mountain hemlock, silver fir, white pine, and subalpine larch. The forest is dissected by frequent avalanche runouts, where trees have been swept away and replaced by lush meadows rich with wildflowers. Still higher, at around six or seven thousand feet in the west, the montane forest thins to treeline, and the ground becomes open alpine country.

In the east, where snowbanks do not linger so long, treeline is higher and in places approaches eight thousand feet. The montane forest below is noticeably more open, and mountain hemlock and silver fir are replaced by Engleman's spruce and lodgepole pine, which are better adapted to the arid climate. Further downslope, where the North Cascades meet the wide valleys of the interior, open ponderosa pine forests, grasslands, and expanses of sagebrush cover the land.

The hike in to the high country is not long, but it is hard. From the small parking lot at the Hidden Lake trailhead, we follow the trail though a thicket of Sitka willow and alder and into a forest of silver fir and western hemlock. Soon, we emerge onto a wide avalanche

chute and began climbing switchbacks. Although weighed down by
eighty pounds of food, books, and mountaineering equipment, we
still can't help but stop and visit the countless wildflowers that bloom
in abundance all around us. It seems as if our timing was perfect, as
if somehow we picked the ideal week for the most different species of
flowers to be blooming at once. Mid-July. We are surrounded by corn
lily, starflower, Solomon's seal, western trillium, white clover, trailing
blackberry, creeping buttercup, bleeding heart, western spring beauty,
wall lettuce, bog blueberry, twinflower, white heather, pink mountain
heather, yellow mountain heather, wild strawberry, and pearly everlasting. The rush of life is almost overwhelming. We take it as an inspiration to keep climbing.

The switchbacks are relentless. In two miles we climb two thousand
feet. Unburdened, we would find this easy, but weighed down as we
are, it is work. As afternoon clouds gather and a light rain begins to
fall, we gather in a small grove of mountain hemlocks for a short, sheltered break. Our route leaves the trail here and heads straight up steep
snowfields to the high alpine pass just north of Hidden Lake Peak.

Our calves are stretched and strained as we step up the snow
toward the horizon line above us. Although the going is hard, the
anticipation of the view to come urges us on. When we arrive on top
of the pass, the North Cascades are laid out before us. We revel in
another completely new panorama of mountains tumbling away to
the horizon as far as the eye can see. We make jokes about how
normal it is becoming to not only see such places firsthand, but to
actually *live* in them awhile. Giddy, we set about the business of making
camp: stamping out tent platforms in the snow, finding water, and
scouting the surrounding area.

At sunset, we get our first tinges of northern light. The sun sinks
below the convoluted horizon not to the west, but to the *north*west.
It blazes peach and crimson behind the silhouettes of the conical
Mount Baker and the serrated backs of British Columbia's Coast
Range. The steep consecutive ridges of the Western Domain turn blue
and dusky, then deep gray, hazier with distance, overlapping, as they

plunge down into the Cascade and Skagit River valleys. To the east the scene is totally different. Here the high peaks of the North Cascades rear their glacier-garnished summits skyward. Scores of white creeks fall in continuous cascades down the green flanks of mountains. They almost free-fall three, four, five thousand vertical feet in a mile. Snow hugs the upper slopes and green and brown heathers are strewn between. But though verdant life clings to the slopes, the high peaks and fluted ridges are all rock and ice, and these, along with the deeply incised surrounding valleys, clearly tell the tale of a time when Cordilleran ice advanced over the entire range, and ice covered the ground from these mountains all the way to the sea.

This place so strongly resembles the Southern Alps of New Zealand that it is uncanny. This is no unexplainable coincidence. The latitude, proximity to the ocean, prevailing weather systems, extent of glaciation, and even the geologic histories of the two ranges mirror each other. Both sit squarely in the latitudinal belt known as the roaring forties. Both are maritime ranges, within view of the sea. Both are barraged by nearly continuous weather systems from the west. Both experienced extensive glaciation throughout the Pleistocene, and still hold significant amounts of ice today. Both formed due to small-scale collisions of microcontinents into larger continental landmasses. The relation between glaciation and bedrock is similar. In both places, the Pleistocene glaciers were so thick and the bedrock so soft that the glaciers were able to cut the valleys nearly to sea level, resulting in astounding local relief and incredibly steep valley walls. Sustained high routes across the mountains are difficult because of these deep incisions in the landscape. As with the Southern Alps, although most of the peaks do not exceed ten thousand feet, the relief from valley floor to mountain top is both steeper and greater than most ranges of fourteen or fifteen thousand feet.

Although the North Cascades sit smack-dab in the middle of the North American Cordillera, the topography here is more similar to the Alps, the Southern Alps, or even the Himalaya than it is to the

Sierra Nevada, Basin Ranges, or the Rockies. In fact, they have often been called the American Alps due to their dramatic local relief, which is undeniably aesthetically similar to that of the Alps of Europe. But to name a mountain range for another is to deny the fact that the North Cascades are a world-class mountain range in their own right. Calling the North Cascades the American Alps is almost like calling the Alps the European Cascades.

Dusk falls at 10:30 P.M. The days are growing longer. I'm starting to actually *feel* the length of the Cordillera, that row of mountains that spans so much latitude. The influence of the North is strong here, beyond subtle. It is in the heaths that surround our camp. It is in the dark, tight-needled conifers on the surrounding slopes. It is in the fifty new species of plants we saw today. It is in the cold, clean water. It is in the glacial ice. It is in the lingering crimson light in the northwestern sky.

Hidden Lake

# VI

With the coming of summer to the alpine zone, we are surrounded by life. Marmots the size of bear cubs frequent the outskirts of camp. They are bigger than those of the Sierra Nevada and sing out in melodic whistles instead of the alarming squawks of their southerly relatives. Pikas, miniature, round, fuzzy, short-eared bunnies, dart among the rocks at the edges of the ridges that rise on either side of us. They, too, have a different dialect from their southern cousins. It is less dry and squeaky, as if the wetness of this place has seeped into them. A chicken-sized ptarmigan hen hides tiny puffball chicks beneath her wings. When she is startled, the young ones explode from her warm, downy underside and scatter. Almost immediately afterward, they start chirping plaintively and making their way back to their mother. I count five chicks, but then a sixth one, even smaller than the others, pops out from behind a rock and runs desperately across the snow on its little legs. It slips once, perhaps still not used to those legs after only a day or so on them. The chick recovers with a bewildered look on its face, and scampers back to the safety of its mother.

The ridge immediately to the north of camp rises steeply and extends northeast. It is rocky, windswept, and green with alpine veg-etation. We call it Scotland because it strongly resembles a small chunk of the Scottish Highlands stuck here in the middle of the greater Cas-cadian landscape. A faint trail leads directly up the ridge and follows the length of its undulating spine to the Triad, three distinct peaks of gray, wet stone, beyond which the ridge becomes obscured by steep gullies and cleavers blocking the view. We head up the ridge to Scot-land, to investigate the alpine vegetation.

The first thing we notice, even before leaving camp, is the abun-dance of heaths in the area. Unlike the Sierra Nevada, where heaths are present but share the ground with other plants, here the heath family dominates the plant community. Most recognizable are the white heather, which also grows in the Sierra, and the pink heather, presumably a close relative of the Sierran red mountain heather.

Among these also grows yellow heather, which is new to us. But heathers are not the only heaths around. They are intermingled with bog and Alaskan blueberries, also new to us on our journey north. All five species are in flower. The last time I identified so many heaths growing in such a small area was in Scotland. Our nickname for the ridge seems appropriate. The feeling of the North surrounds us.

We make our way up to Scotland, but progress is slow because there are new plants everywhere. Clinging to the rocks is an unfamiliar species of wild strawberry. Moss campion hugs the exposed ground in tight cushions. Common juniper forms close shrubby mats around the protruding rock knobs. All three of these species occur throughout the boreal or tundra regions of the circumpolar North. But plants with alpine origins are here, too. The snapdragon family is represented by three species new to us, including Cusick's speedwell, an unfamiliar species of red paintbrush clinging precariously to the rocks, and alpine penstemon, replacing the conspicuously missing Sierra penstemon. We find two species in common with the Sierra Nevada, including Davidson's penstemon, and showy polemonium of the phlox family. There are no buckwheats here to speak of. They have been largely left behind somewhere in the Southern Cascades.

The Cascades, as common sense suggests, have significantly more plants in common with the North than do the Sierra Nevada, because the Cascades are closer to the North, more directly connected to the North by contiguous migration pathways, and because the climate of the Cascades emulates that of the North more closely. Heaths, for example, are diverse and abundant here, forming communities unto themselves where acidic conditions prevail. A diverse assemblage of willows lines the streams above and below treeline. Among the family Pinaceae, spruces dominate the forests of the interior, accompanied by lodgepole pines, both of which extend northward throughout the greater boreal forest. On the west slope, mountain hemlocks form the biogeographic link with the north, extending along the maritime mountain front well into Alaska. But we are still in the mid-latitudes, still just halfway up the Cordillera from its southernmost alpine outpost. The beautiful purple shades of Davidson's penstemon and showy polemonium are here to remind us of that.

After a few hours of botanizing, we spend some time surveying the upper ridge. Tomorrow we plan to traverse it in hopes of making it over to the Eldorado Glacier. The glacier itself is only a few miles away, but the terrain between here and there is intimidating to say the least, and we are uncertain of our passage. The only other way up to the glacier is to head back down to the Cascade River valley and ascend a notoriously brutal approach trail up Eldorado Creek, which ascends over five thousand vertical feet in three or four miles to reach the edge of the ice. The Eldorado Creek route is well traveled, and is more of a sure thing. But it is also more of a surely painful thing. I have never met anyone who has tried the ridge traverse. It promises more adventure, and more elegance.

The first person known to pioneer a traveling route across between Hidden Lake Peak Pass and the Eldorado Glacier was the legendary Fred Becky, whose name has become as synonymous with the Cascades as John Muir's or Norman Clyde's is with the Sierra Nevada. Becky is known to understate the difficulty of off-trail travel in the High Cascades, and provides only brief descriptions of his routes that leave room for a healthy amount of adventure. Becky's description of the route says little more than follow the Triad ridge over towards the Eldorado glacier. He makes it known that such a traverse is possible but gives little else away. If we can make it, it will take us ten hours of traversing steep ridges and snowfields, rappelling choked gullies, skirting rockbands, climbing boulder chutes, and roped glacier travel to get us to Eldorado. But every minute will be worth it.

# VII

It takes us ten and a half hours, and we reach the Eldorado Glacier in time for supper. We are tired and spent after a long day on rugged terrain but feeling good. To have attempted the Becky traverse any earlier in the course progression would have been disastrous, but our hard work in the mountains has paid off and our skills and level of fitness matched the task at hand. We make camp on the ice and watch the evening show of light.

The glacial plateau, like so many other places in the Cascades, looks and feels just like the Southern Alps of New Zealand. The black, glacier-covered heap of Eldorado Peak rising out of the ice to our north could just as well be Mount Aspiring, and the plateau itself the Bonar. The Cascade River Valley to the south is a northern reflection of the West Matukituki Valley, and the bold form of Johannesburg Peak across the valley just like Mount Barf of the Southern Alps. I shake my head in disbelief, but the similarities are real and they are many, and they are so similar that I am stunned.

The plateau is big by lower forty-eight standards—over ten square miles if taken with all of its interconnected parts, including the Inspiration Glacier and the Klawatti system. The surface texture of the ice is mostly predictable. Where the ice is relatively flat it is benign, its surface smooth and white, easy to travel over. Where steeply undulating bedrock causes the ice to bend too much, it cracks, forming yawning crevasses and blue icefalls. There are undoubtedly a few surprises hidden beneath the thin layer of snow that still covers the bare ice. Dark gray peaks rise up out of the glacier like great teeth, small nunataks, stark islands of stone inundated with ice. Of these, Eldorado is the largest and the most prominent but the least dramatic in profile. It looks like a tilted block with an enormous snow cone on top. From our camp, the snow cone looks like a whale's back. It appears low angle and looks to be an easy climb. But it is rather a fin, sculpted and sheared by the wind to a fine edge that drops off at angles in excess of fifty degrees on either side. Our view is sidelong and comforting but ultimately an illusion.

The plateau is now an isolated island of ice, once connected to the greater mountain icecap of the North Cascades, and during times of maximum ice connected to the entire Cordilleran Ice Sheet. The ice of the Eldorado Glacier in particular used to flow into the Cascade River Valley and out the mountain front. Today, however, the once-giant valley glacier of the Cascade River Valley is completely gone, and without a greater stream of ice to feed into, the Eldorado Glacier, like all the surrounding glaciers, is cut off, disconnected, isolated, removed, suspended, a castaway. At its edges, it is broken into a thousand blue, swirled blocks that periodically detach in car-size chunks and trundle violently downslope into oblivion. But even as the land of the living below slowly consumes the ice, the ice exacts a price among those of the living. Misguided butterflies lie limp on the surface of the glacier, blown up here by the wind, chilled by the ice to the point of immobility. I pick one up gently, hold it in my cupped hands and breathe warm air over it. After about a minute, the butterfly starts to stir. Eventually it flutters away, and I wonder if, given this chance, it will make it off the ice.

The evening sky is, as has become usual, spectacular. The setting sun sinks below the horizon and the Coast Ranges of British Columbia and the Olympics stand silhouetted in a soft, orange light. The light spills out from behind the horizon line, illuminating the thin layer of scattering clouds from beneath, bathing the mountains and glaciers in a subtle peach alpenglow. This is a beautiful, clean, uncomplicated world.

In the morning, we are enveloped in clouds and for two days are suspended in a world of white. Rain drives down from an invisible sky and turns the surface of the glacier to a slushy mess. The wind sweeps across the plateau like a drunken banshee, pushing and pulling at our limp, soggy tents. The empty sense of living in a flat, white, cold world is pervasive. We sleep. We melt snow and make hot tea. We read. We emerge between rain showers and run laps around camp. We gather in huddles, share a few jokes, get cold; the rain starts again, and we retreat. When the rain stops for more than an hour, we suit up and practice our glacier systems on the wet ice, and rope up for blind forays on the plateau to practice route-finding and have a look at the insides of crevasses. But Eldorado Peak is completely gone, and with it the idea of climbing. July twentieth. Forty-one degrees Fahrenheit. Summer has abandoned us.

Ultimately, the weather affords us more time around camp, and we make use of it by discussing the geologic story of the North Cascades. It is a fascinating history because in many ways, even moreso than the story of the Sierra Nevada, the geologic history of the North Cascades is a microcosm of that of the entire Cordillera. Here, all the dynamics of accretion of new lands, metamorphism of continental crust, subduction of oceanic plates, arc volcanism, fluvial erosion, and glaciation that typify the Cordillera have been and currently are at work.

All of the many geologic processes at work in the evolution of the North Cascades, and the entire American Cordillera for that matter, are associated with the subduction of oceanic plates beneath the overriding continents. The subducting oceanic plates not only cause extensive volcanic arcs and associated granitic basements to form, but also act as conveyor belts that continuously deliver new terranes to the edge of the continents. These new terranes may be maverick pieces of continental crust, sometimes called microcontinents, or scraped-up shards of sea floors that get pasted into the mix. The classic arc volcanism and emplacement of granitic basements result in a landscape that looks just like the Southern and Central Cascades do today, and the way the pre-Sierran arc volcanoes likely looked some hundred million years ago. Add a complicated jumble of accreted terranes into the mix, and you get the North Cascades, the Canadian Coast Ranges, the Icefield Ranges, and most of Alaska.

Geologists recognize three major terranes that converged to form the bulk of the North Cascades. Each of these terranes collided with the western edge of North America at a different time and is bounded by significant faults on either side. Each of the three terranes is made up of similar rock types, but in no way is each terrane a uniform unit unto itself. Rather, each terrane is an amalgamation of still smaller terranes and scraped-up seafloor wedges that are sandwiched in between, and these smaller component parts may differ in age from one another by hundreds of millions of years.

Each time a terrane was rafted up against the continent, a small-scale version of an Alpine-Himalayan orogeny occurred. The arc volcanoes associated with the subduction of the intervening seafloor built up and were eventually sandwiched between the colliding continental crust. Once the seafloor was devoured, the remaining shards of it that had been scraped up onto the leading edges of the encroaching continents were pushed together and thrust upward by the collision. As the action continued, pressure caused the varied rocks to metamorphose, to different extents in different places and at different times. Pressure also deformed the rocks into folds and nappes, and heaved the crust upward into a complex system of mountains.

The first of the three major terranes to collide with North America is known as the Methow Domain. It comprises the easternmost part of the North Cascades. It is dominated by unmetamorphosed seafloor sediments that accumulated at the edge of the continent as mountain-building processes were just getting under way. Various volcanic rocks indicative of early stages of arc volcanism are found scattered throughout the domain. Since the rocks of the Methow Domain are relatively soft, they do not form the dramatic mountain landscapes that typify the North Cascades but, rather, broad plateaus and shallow valleys.

Next came the Metamorphic Core Domain. As its name suggests, this terrane is the centerpiece of the North Cascades and more than any of the others typifies the character of the range. All the range between Ross Lake and Washington Pass to the east, and Marblemount to the west, including the Picket Range, Eldorado Peak, Liberty Bell, Dome Peak, and the Stehekan Valley are within the Metamorphic Core Domain. All types of rocks are represented here, most of them heavily metamorphosed and many of them impressively resistant to erosion. These already tough rocks were later reinforced with granite intrusions, which added significantly to their rugged character.

The last of the three terranes to arrive was the Western Domain. It is recognized as the most complex and convoluted of the three major terranes and is itself a folded stack of smaller terranes smashed together. The Western Domain finds its most exaggerated relief in the striking form of Mount Shuksan but typically has less dramatic topography than the Metamorphic Core.

By ninety million years ago, about the time that arc volcanism in the Sierra region was winding down and erosion was about to set in, all three of the major terranes of the North Cascades were in place and things quieted down a bit in the region. The work of erosion became preeminent, and the mountains slowly but surely wore away.

The North Cascades were revitalized by large-scale Cascadian volcanism that ensued around thirty-five million years ago and continues even to the present. As was the case with the pre-Sierran arc volcanoes, plutons of magma distilled from the mantle welled up and intruded into the overlying crust. As these plutons worked their way through the terranes of the North Cascades, they caused a further metamorphosis of the crust, welding it together and toughening it even more. Where plutons reached the surface, huge stratovolcanoes built up, much like Mount Baker and Glacier Peak of today. Where the plutons didn't make it out, they solidified into granite beneath the cover of older metamorphic rocks. As time went on, the oldest of these volcanoes completely eroded, but their granite basements remain as evidence of their previous existence. The present-day large volcanoes, most notably Mount Baker and Glacier Peak, are relatively young additions to the range and signify the continuing importance of Cascadian volcanism in the shaping of the North Cascades mountain landscape. Baker, for example, is less than thirty thousand years old. Ten-thousand-year-old lavas encrust its cone.

In the last three million years, glaciers have wreaked havoc on the

North Cascades, sharpening summits, scooping out valleys, and shaping innumerable bowls, cols, and arêtes. The overall effect, as in the Sierra Nevada to the south, has been a dramatic increase in vertical relief in the range. Here in the North Cascades, however, two important details differ from the Sierra, which together explain why the North Cascades make the Sierra Nevada seem like easygoing terrain. First, the mountains here were more rugged to begin with. While the Sierra were uplifted as a long, contiguous fault block, the North Cascades were raised as fold-nappe mountains, inherently convoluted from their conception. Second, the rock of the Sierra is almost entirely granitic and is uniformly resistant to the effects of glaciation. The North Cascades, however, are made up of a mosaic of rocks that are in no way uniform, most of which are more susceptible to the effects of glaciation. The result is more dramatically dissected terrain: jagged peaks, torn-up ridges, and deeply carved valleys. These characteristics are naturally most pronounced in the west, where glaciers have always been the most extensive. To the east, where the rain shadow effect moderates and eventually checks the extent of glaciation, erosion is primarily by water.

All this talk of mountain building and glaciation brings our attention back to Eldorado. We look out into the white void, and the thick veil of cloud makes it seem as if the world is still in the throes of creation, as if the mountains are rising from the steaming bowels of the earth even as we sit, jaws gaping, staring into what we know cannot be nothingness. Eldorado is in there somewhere, the mother of the Metamorphic Core Domain, and it is calling us. Today is our last day on the mountain, our last chance to attempt a climb of its snowy flanks. After some discussion, we decide to make a go of it.

At 4:00 P.M., we rope up and set out into the white blankness, heading north in the direction of Eldorado. The way is flat and the ice benign except for a few obvious narrow cracks that seem to run almost the full width of the glacier. In the cloud, we miss our intended entry ramp onto the Eldorado ridge by several hundred feet.

Instead of staying on the snow, we ascend to a small notch in an exposed rock ridge. We travel over the rock and it leads us onto the ice of the ridge. Here we stop to put on crampons, organize pickets and screws, and make any last-minute preparations necessary before the ascent.

We climb swiftly at first; the route poses few hazards, so we go unprotected. After crossing between several small crevasses to our right and a precipitous rock band to our left, we begin setting snow pickets as running belays, so that each rope team is clipped into at least one picket at all times. As the snow steepens, our attention is increasingly absorbed by the mountain. Stepping upward, the snow gives way beneath me and I find myself falling. Before I have time to be scared I stop, wedged thigh deep into a crevasse. I carefully pull myself out, then clear the thin snow bridge off the surface of the ice to reveal the crack. It is only two feet wide, but plenty deep. Nevertheless, with its edges clearly defined, we can jump over it without difficulty. We protect the jump by setting a screw in the bare ice above, and continue climbing.

The ramp of snow and ice takes us to the narrow snow fin that leads to the summit. The top of the fin is only a foot or two wide, and made so only by our stamping it out. The angle of the ridge is unquestionably acute, and it drops off on either side into a swirling, white, bottomless abyss. We follow the fin in careful single file, specks moving silently, slowly, deliberately, sifting through the clouds, making our way along the white spine of the mountain to its shrouded top.

The angle of the ridge lessens and grows flat, and we follow it to where it gently slopes down to the dark shapes of rocks below. It is the summit, though we would not know it if the world didn't drop off on all sides. Visibility is twenty feet. It is an odd feeling, being perched there on top of one of the highest peaks of the North Cascades and feeling like we are floating in the middle of the primordial void.

Cold and becoming wet, we do not linger. We soon find ourselves descending the narrow fin, tiptoeing a second time across that ethereal white bridge, and onto the steep, sloping ramp below. Because we left our pickets in place, the descent goes quickly, and we move much faster than on the way up. In less than an hour we are all safely back at the snow notch, shedding crampons and huddling up for thanks and smiles. Astonishingly and without notice, the sky opens overhead, illuminating the icefalls of the Inspiration Glacier below and the stark, ice-encrusted lower wall of Forbidden Peak to the east. The sudden burst of light revives us, and it is as if we have awakened to the light of the world for the first time. It gives context to where we are and what we have just done. But after a few precious minutes, it is gone, and we set out into the cold, white void, awaiting its return.

Eldorado Glacier

Descent in whiteout

# VIII

One hundred miles north of the Canadian border we lose the hum of cities and even from the air there is scarcely a sign of civilization in sight. We have entered the world of wilderness that lies beyond the fiftieth parallel. A hundred thousand square miles of mountain wilderness. Two hundred thousand. Maybe a million square miles or more. From the air it looks infinite. The sky is the color of a bluebird and as clear as a bell, and the earth reveals hues of soft green and blue, streaked with white and gray. My face is pressed up against the airplane window. Maps may give names to the many ranges that make up the Cordillera, but from thirty thousand feet up they appear as one great continuous mountain landscape that stretches infinitely to the north and south. The North Cascades blend almost imperceptibly into the Coast Ranges of British Columbia, and the Cascadian themes of radical relief and heavy glaciation intensify. The mountains rear their shoulders ever higher above the trees, and as we go north their recesses are filled, first to the brim, then overflowing with swarming, riverine glacial ice that threatens to spill out across the whole world. The scene culminates in the massive form of Mount Waddington, so heavily encrusted with ice it appears as a gigantic glacial mound, heaped and broken, weighed down by its immense glacial burden. The mountains grow, the tundra grows, the ice grows, the green spaces between the mountains shrink, the lights go out. We are going North.

To the west, the mountains fall abruptly to the sea, and the coastline is dissected by an elaborate system of deep blue fjords. From the air, the concept of mountain valleys flooded by rising sea levels is pictured clearly and perfectly. Fingers of wrinkled land reach far out to sea. Fingers of sea reach far into the land. The coastline looks so much like that of western Norway that if I didn't know better I would insist I was there. I would only be able to differentiate between the two after time on the ground, after having a look at the plants. If I were up on the tundra, or on the ice, I could wander for weeks and not know British Columbia from Scandinavia.

To the east, the mountains become lower, glaciers exponentially smaller, fjords and lakes become absent, and the green hue of the forests becomes more faded and less deep. Although the country is less rugged, it is more expansive, open, and unbounded by the sea. It is the broad sweep of the interior, the western edge of the great boreal forest that stretches east clear to the coast of Labrador. In the distance, far off on the eastern horizon, it appears as a hazy blue line, like a sea in both scale and color, but a sea of trees.

The North is both easy and complicated to define. A simple definition would be to take out a range map of white spruce, the most common and widespread species of the boreal forest, and describe the North as everything above the southern edge of the spruce's range. Interestingly, you would find that a score of plants and animals that are considered characteristic of the North, maybe even a hundred other species, have ranges that match up pretty closely. But is it really that simple?

Human perspectives in particular are complicated. To a Mexican, the American Southwest is the north. To a Southern Californian, San Francisco Bay might be considered the north. Canadians from the population centers of Vancouver, Calgary, Edmonton, and the various cities of Ontario and Quebec might consider the Yukon Territory, Northwest Territories, Nunavut, and Labrador the north, while to those living above the Arctic Circle the bulk of the land in those regions would be considered south. Physical factors and biogeographic perspectives (and their inherent effects on human cultures) provide easier definitions. Most geographers agree that the greater North consists of the Subarctic, where trees are present, and the Arctic, where they are not. This sounds simple enough, but it is not. Nature is far more subtle, flexible, creative, and complicated than that.

The Subarctic coincides with the boreal forest biome, characterized by nearly contiguous conifer forests, long, cold winters, and short summers. The southern boundary of the Subarctic is defined differently by different people. The most liberal definition includes the broad transitional belt between the deciduous hardwood forests to the south and the definitively boreal forests to the north, known to many as the North Woods, and characterized by the presence of maples, beeches, and occasional northern red oaks, in addition to the spruces, pines, firs, and associated birch and aspen of the greater boreal forest. Other more conservative definitions draw the line farther north, and strictly exclude the North Woods from the boreal because of the presence of "southern" hardwoods. Another biological boundary that helps define the southern edge of the boreal is what the veteran northern studies biologist Steve Young calls the "vine line," the northern extent at which a variety of plants, including honeysuckles, grapes, and poison ivy, can grow. The "vine line" provides a meaningful boundary because many species, not just one, have their range in common. The "vine line" is inclusive of the North Woods–type forests.

The northern boundary of the boreal is easier because it coincides with a decisive climatic factor: the Arctic treeline. Simply put, trees can grow only where the summer (July) growing season temperatures average fifty degrees Fahrenheit or more. South of this "line" is the boreal forest, and north of it is the Arctic tundra. In reality, however, the Arctic "treeline" is no line at all, but rather a broad transitional belt (not unlike the North Woods in this respect) a score to two hundred miles wide, in which the boreal forest opens up, and its trees become stunted and gradually disappear. Swaths of pure tundra reach deep into the boreal forest, while patches of sheltered boreal forest extend north onto the sea of tundra like peninsulas and islands of trees.

North of the treeline, the Arctic tundra extends north to the Arctic Ocean, and continues north across the smattering of islands that rim the Arctic (including Greenland, the largest island in the world) even beyond eighty degrees north latitude. The boundaries of the tundra, like the landscape, are clean and relatively simple. The tundra biome is bounded by physical factors on all sides: treeline (the fifty-degree-Fahrenheit isotherm) to the south, and the absolute limit of plant growth to the north. Beyond the greater Arctic tundra, the discussions of biomes come to an end.

Add mountains to the picture and defining the North gets complicated. Because the altitude of mountains simulates latitude, fingers of more northerly biomes grope south and confuse simple latitudinal models. Nowhere is this more subtle, and thus confusing, than in the North itself, where a broad mountain plateau might rise a thousand feet above a boreal forest floor and result in a raised bed of tundra that sweeps across the landscape for a hundred miles. Is it alpine or Arctic tundra? Does the presence of a few scattered trees make it boreal forest, or is it treeline, and if so, which kind, alpine or Arctic? Then add big glaciers. At sixty degrees north latitude, are they polar or non-polar? If they originate in the high mountains but flow clear down to the sea, are they alpine glaciers or not? Where they have melted off and nothing grows, is it tundra, boreal forest waiting to happen, or something all its own? Human definitions and categorizations for nature become challenged, stretched, and torn. We are reminded that nature is far more subtle, flexible, creative, and complicated than we are capable of understanding.

Definitions of the North aside, north is where we are going, in the direction of Polaris by night, following the red needle of the compass and adjusting for declination by day. Both the landscape below and the feeling inside me confirm that.

The juxtaposition of the fjord country to the west and the interior to the east is striking. Both views typify the North, but in different ways. Fjord country occurs in subpolar latitudes where continental landmasses rear up mountains along the coast in the face of prevailing weather systems. The maritime influence of the close proximity of the oceans means two things: lots of moisture and moderate temperatures. Precipitation comes in loads, in the form of rain, sleet, hail, and snow. All this moisture amounts to the formation of glacial ice in the mountains, and so much precipitation means that the abundance of glacial ice may even flow down to elevations where it would never form on its own. If conditions are right, it might even reach the sea. During the Pleistocene, the glaciers were so big they all reached the sea and far out into it, carving valleys deeper and deeper as they went.

When the ice melted, sea levels rose and filled the emptied glacial valleys with saltwater, giving the country the appearance of a flooded
mountain range, which is exactly what it is. The maritime influence
also means moderate temperatures, which means a place has to be suf-
ficiently far north and/or high enough for snow to fall despite the
moderating effects of the ocean. It means no fjord country in
northern California, Oregon, or southern Washington, all of which
are continually washed in rain but rarely see snow in abundance.

With some basic knowledge of physical geography it is pretty easy
to predict where fjord country should be on the planet. It should be
north or south of around forty-five or fifty degrees latitude. It should
be in areas where there are mountains, especially where there are big
mountains. It should be in places that face west, though east will do if
the place is far enough north or south. Trace your finger around the
globe and these places will come up: coastal Alaska and British
Columbia, the entire fringe of Greenland, the west coast of Norway,
the west coast of Patagonia, southwestern New Zealand. These are the
places on earth where fjord country attains its greatest glory. But there
are other places that are more subtle, where the landscape is less dramatic in vertical relief but no less dissected: coastal Labrador and Newfoundland, the Maine coast, Iceland, the west coasts of Ireland and
Scotland. In all of these places the glaciers of the Pleistocene ground
deep gouges into the land to be filled in by the sea.

The interior is different in all respects save one: it is still a cold
place, and it is cold that defines the opposing worlds of both the coast
and the interior as definitively northern. The interior is begotten by
the rainshadow effect of the Cordilleran mountains, and is the
northern analog of the Great Basin desert and open pine forests of the
east side of the Sierra-Cascade axis to the south. It is a continental
region, and though on a map it may seem close to the ocean, the
intervening mountains exaggerate the degree of continentality of the
interior to make it seem thousands of miles from the nearest coast.
Precipitation is relatively low, in some places approaching desert conditions. Snowpack is thin and dry compared to the thick, heavy cloak

of white that mantles the coastal regions. Temperatures are extreme, soaring to the nineties in summer and regularly plummeting to fifty below in winter. Glaciers do not form here; there is not enough moisture to support them. Without the dramatic sculpting and scouring of glaciers, the landscape is eroded by water and wind, and is relatively broad and undulating. The expanse seems endless, and it almost is. It sweeps across nearly all of Canada almost to the coasts. It picks up again in eastern Scandinavia and spans all of northern Eurasia, twice the girth of North America. It circumscribes the landmasses of the Northern Hemisphere. Geographic knowledge will also make it easy to predict where interior boreal country will occur. It will be inland from fjord country, on any continental landmass higher than forty-five or fifty degrees latitude. In the Northern Hemisphere such places are common. In the Southern Hemisphere there are none.

Geographic maps of the North American portion of the North reveal some interesting coincidences in climate, biogeography, and cultural geography. The climate map shows three or four zones. To the north and hugging the exposed coldwater coastlines of Western Alaska, Hudson Bay, and Newfoundland, is the Arctic zone, characterized by July temperatures averaging less than fifty degrees Fahrenheit. Below and within this zone is the Subarctic zone, taking up most of the map. Some maps might further divide this into the Eastern Subarctic, relatively well-watered with dense forests, and the Western Subarctic, relatively dry with open forests. The smallest zone is the Pacific Coast zone, which follows the coastline Southeastern Alaska to British Columbia, bounded on the west by ocean, and on the east by the mountains of the Cordillera. The biogeographic map also shows three or four zones: Arctic tundra, boreal forest (further subdivided into east and west), and Pacific temperate rainforest. All of these biogeographic zones match up perfectly with the climatic zones, and this should come as no surprise, because climate is the ultimate arbiter of the distribution of natural communities. What is amazing is that the cultural map also coincides, almost perfectly, with the natural boundaries of climate and biota. The ancestors of those early Americans who followed the coastline in their migrations, including the Haida and Tlingit, among others, now occupy the Pacific coast regions. The migrants who followed the ice-free corridor of Beringia came in waves. The Algonquin-speaking peoples, including the Cree, Ojibwa, and Wabanaki, among others, made it to the eastern boreal and are still there today. Later waves of Athabascan-speaking peoples (Dine), perhaps related to the Haida and Tlingit, came later and for the most part remain in the western boreal. Finally, successive waves of migrants, culminating in the Eskimo-Aleut-speaking peoples, developed the technologies necessary for living on the open tundra and moved into the Arctic. Each of these cultures has traditionally been so specifically adapted to its landscape as to be virtually inseparable from it. Where there is Arctic tundra and open coast there are Eskimos and Aleuts. Where there is boreal forest there are Athbascans in the west and Algonquins in the east. Where there are tall trees and salmon there are Haida, Tlingit, and Eyak.

The Icefield Ranges of the Alaska-Yukon border country straddle the sixtieth parallel and occupy chunks of Alaska, the Yukon Territory, and the province of British Columbia. These ranges belong as much to the North as they do to the Cordillera, and it is here that the characteristics of both of these geographic regions mix and intermingle. This is a mountain landscape of huge proportions, a crown jewel in the Cordillera that sits squarely in the North. The Cordilleran story of subduction zones, accreted terranes, arc volcanism, Pleistocene glaciation, and species migrations reaches new levels of intensity here, and is made more rich by the inclusion of all the elements of the North, including coastal and interior boreal forests, muskeg, treeline, alpine tundra, extensive glaciers, healthy populations of large mammals, and a rich human history. They are called the Icefield Ranges because they are home to the largest nonpolar icefields on Earth, and it is this ice that gives common identity to otherwise separate ranges. The Saint Elias Range forms the nucleus of the group, with Mount Logan as the centerpiece, the source from which much of the ice of the icefields flows. The Wrangells to the north and the Chugach to the south also contribute a share of ice, as well as several smaller subranges. All told, the Icefield Ranges span over four hundred miles in length and over a hundred miles in girth. But their boundaries are blurred. On a topographic map they blend imperceptibly with the greater Chugach and Alaska Ranges to the west and the Coast Ranges to the south, which themselves are links to the Aleutians in one direction and the rest of the American Cordillera in the other. They are a link in the chain, but a big, heavy, bulky link, covered almost completely with ice. Here, amid the ice and gravel-strewn rivers and grizzly bears, is where we go to bring completion to our Cordilleran journey.

North of Mount Waddington, we follow the clustered heaps of mountains north. The summits and glaciers and lakes and fjords seem beyond number, and I give up on trying to count them all. Somewhere above the fifty-fifth parallel, we veer noticeably to the east, away from the coast, and into the interior. We will not see the Icefield Ranges today. They will remain a mystery until another time. Now, the smooth contours of the interior open up. Somewhere out there, farther east than I can see, was once the ice-free corridor, that place where migratory ground sloths, short-faced bears, antelope, steppe bison, and eventually humans slipped between the hulking masses of the Cordilleran and Laurentian Ice Sheets and found passage south, to emerge in the sun-washed Eden of the Serengeti-like Great Plains of North America.

The Yukon Territory. Some time in the last hour we passed into its skies. Tucked up against Alaska, the Yukon has somehow escaped the publicity that its grandiose neighbor to the west has attracted. Bookstores across the continent have entire bookcases dedicated to Alaska but seldom a single shelf of books on the Yukon. Most Americans' idea of the Yukon is the tiny strip of it seen from the road on the way to Alaska, and perhaps some vague recollection of a gold rush. In fact, it dwarfs forty-eight of the fifty states, but has a population ten times smaller than the least populated state. Two-thirds of the Yukon's fifty thousand live in one town, the territorial capital, Whitehorse, on the banks of the turquoise-blue Yukon River, mightiest of all the rivers of the North, whose headwaters are found in the blue ice of the Icefield Ranges.

Below us, a conspicuous blue strip of river cuts cleanly through the dull green of the boreal forest. The banks of the river are lined with bright green thickets of alder and willow, and somewhere in those thickets are bears—big ones. As we descend, I can see that the waters are clear, cold, and fast. It is the mighty Yukon, and it is flowing north. We are going down there.

# IX

We are surrounded by fluttering aspen, bright magenta fireweed, and dusky white spruce. The air is hot and dry, the ground dusty, the understory vegetation parched and crackly. This place reminds me of the slopes of the Rockies, where aspen and spruce continue their interior boreal association far to the south. But this is no southern extension. We are in the heart of it. Members of the pine, willow, and heath families define this place, detailed by the presence of an assortment of roses and birches. Where meadows meet the forest, avens have gone to seed, all of them at once, covering the open ground with thousands of inch-wide silver pom-poms. Beneath the trees, bearberry covers the ground, along with twinflower and prickly rose, all of them circumpolar species, all of them heaths and roses. The sound of the Duke River roars behind us. Ahead, an old path climbs up through the boreal forest.

I have never seen a forest so dry. Here, near the north end of Kluane Lake, the rain shadow cast by the Icefield Ranges has the strongest effect, which makes perfect sense, because between this place and the Pacific Ocean are the biggest mountains in the Icefield Ranges, including Mount Logan, the highest in all of Canada. Precipitation levels are less than thirteen inches a year. Even in the heart of winter, snow is usually but a dusting, and rarely amounts to anything one would consider snowpack. Fire is a regular occurence, clearing out the understory at regular intervals or, if suppressed for too long, burning the whole forest to the ground.

We ascend the path through the forest and it climbs steeply. Soon, we are flanked on either side by closely growing ranks of willows. We begin clapping our hands, making noise, singing songs out loud. Bears could be in there. Already we have seen tracks, down by the river, in the mud.

After only a few miles, we emerge from the trees and onto the befuddling vastness that is the muskeg tundra. It is a plain of rolling prostrate vegetation growing on an underlying cushion of moss. There are few trees, and those that are out there are spindly, dwarfed, undernourished versions of their forest counterparts. They look like thrashed bottlebrushes perched awkwardly on the otherwise open landscape, like tall buoys floating on a sea of green. Without them, all concept of distance would be obscured. They give us a scale to work with. The scale is huge.

We set out overland across the springy, uneven mat of the muskeg. It feels like walking across an endless sponge two feet thick with holes riddled throughout it. Plus, we are bearing heavy packs. We waddle our way along, slowly, hard pressed to travel a mile and a half per hour. Though the ground appears flat on our topographic map, the reality of its surface texture is a different story. It has microtopography more radical than the Alps or the Cascades. Our slow progress adds to the feeling of vastness of the landscape. But we are consoled by the fact that we are not knee-deep in water and swarmed by black clouds of mosquitoes. It is late July, and the muskeg here is dry.

While the scale of the larger landscape is huge, that of the flora is miniature. I bend down and look closely at the ground; it looks like a tiny forest. There are mosses and lichens and herbs and shrubs and wildflowers and fruits of all shades of green, brown, gray, blue, white, and pink. But all of this is conspicuously underlaid with a thick layer of moss. It all starts with cold water, either from the sky or from seasonal melt-off of permafrost within the ground itself. Cold, standing water slows decomposition rates to a minimum, and organic material starts to build up. As it builds up but doesn't decompose, acidity increases and exacerbates the situation. Eventually, undecomposed organic material, called peat, forms a blanket over the underlying soil and bedrock. Meanwhile, sphagnum mosses, able to grow in the most acidic conditions and even increase acidity, slowly gain a monopoly and cover the ground. Only those plants able to grow on the sphagnum can gain a foothold in the muskeg. These plants must be able to handle acidic conditions, and they derive nearly all of their moisture and nutrients from rainwater collected in the sphagnum or from the sphagnum itself. Of all the plants that meet these criteria, the heaths reign supreme. All around us and beneath our feet, growing from the sphagnum, are white heather, lingonberry, cloudberry, twinflower, and blueberry— all heaths. These are joined by crowberry, bog asphodel, and dwarf birch. I have heard stories of grizzlies roaming the muskeg in spring, gorging themselves on last year's heath berries rich in fermented sugars, the bears stumbling along like drunkards with bewildered expressions on their faces. As we walk, we reach down and grab handfuls of blueberries and eat them along the way.

We amble along, laughing at each other's blue-stained teeth, mouths, and cheeks. Ahead, starkly visible in the wide-open sea of green, a lone caribou raises its head high above the floor of the muskeg. It is maybe two hundred feet away, upwind, but still it has had no trouble spotting us in such an open space. At first, the sight of a single large animal in the midst of such a large piece of land seems odd, but after some thought it makes sense. In this part of the North the caribou do not form the seasonal herds of tens or hundreds of thousands that they are known for farther to the north and east. Even if they did, July is not the time for that. Here, in midsummer, at sixty degrees north latitude in the interior of the Yukon Territory, the caribou are most often solitary or in small groups. They do not join the latitudinal migration between the boreal forests and the Arctic tundra that their northern and eastern relatives do, but rather take advantage of the local convenience of elevational migrations instead. Why go all the way to the Arctic and back again when there is boreal forest and tundra locally? We watch the animal intently. It watches us. It becomes a staring contest. Eventually, we break our gaze and resume walking over the muskeg. It does the same, and trots off to the north.

We pass the broad hill labeled Burwash Uplands on our map and aim for a subtle pass to the west that will drop us into the gravelly bottom of Burwash Creek. As we travel over the muskeg, we begin to fantasize about the upcoming firm ground of the creek bottom. Our ankles and hips are worn out, muscles strained from so much twisting and gyrating. But though the pass appears close at hand, it is still two hours off. We waddle on.

At Burwash Creek, the scenery changes abruptly for the first time in many miles. The drainage cuts through the surrounding topography and provides a window open toward the glaciated flanks of Hoge Peak. It is our first look up at the high mountains, and we are taken by their grandeur. They are but foothills compared to the mounts that rise to the west, deeper in their ranks. But they are majestic, their snow-clad sides glinting in the sunlight. Closer at hand, the creekbed itself is wide and made of shifting gravel beds. The creek is braided in places, consolidated into a single strand of water in others, and changes its complex course yearly, seasonally, even daily. It is fringed with willow thickets and soapberry shrubs. Soapberry is here because it can grow only in high-nutrient, nonacidic conditions, and creekbeds are some of the only places in the North that fit the bill. Here, the fluvial action of the river continually flushes the place out and deposits fresh, nutrient-rich sediments. Growing in such nutrient-rich places, the soapberries themselves are a rich food source, and attract animals. Most notable of these is *Ursus arctos*, the grizzly bear, whose favorite food is none other than the soapberry. In short, where there is a lot of soapberry there are a lot of grizzly bears. Soapberry is ubiquitous along the numerous watercourses of the interior of the Icefield Ranges, thus Kluane National Park and adjacent areas have the highest density of grizzlies on the continent. We clap nervously, make noise, sing songs. We pass the shed antlers of moose, caribou, and even deer. Dozens of ptarmigan flush before us, clucking in the willows.

The creek is big, swollen with its mix of glacial meltwater from above and recent rainfall runoff from all directions. We follow its course, inspecting palm-sized wolf tracks in the river mud along the way. Then, across the river, a huge, lumbering shape moves through the willows. Leaves shake and limbs snap, and the unmistakable form of an adult grizzly emerges from the bush. At first she does not notice us, unable to hear our racket over the pervasive din of the river. But then she catches our scent. Abruptly, she turns to face us, rears up on her hinds, and nods her head from side to side sniffing the air. As she moves, her grizzled blonde hair cannot hide the huge, knotted muscles that ripple over her frame. They are taut; she is tense. I look to the brush behind her and find the source of her tension. Two yearling cubs, themselves the size of full-grown black bears, run clumsily through the shrubs and up the opposite bank. The sow drops down to all fours for a brief second, then rears up on her hinds a second time, as if still uncertain of our identity or intentions. Then, satisfied, she follows her cubs up and over the bank and disappears into the tundra.

We make camp on an open terrace on the west bank of the creek, within view of the Burwash glacier and the desert-like gorge of the upper Burwash Creek. We set up our kitchen area far out on a gravel bar, a few hundred feet from camp, the memory of the grizzly sow and cubs fresh in our minds. We busy ourselves with cooking supper, talking among ourselves and admiring the surrounding landscape, reflecting on the experiences of the day, reveling in the presence of so many large mammals. As if on command, Granny points out a large red fox that has trotted into our camp. She sits down calmly next to one of our tents, her white and orange coat fluffing in the evening breeze, and proceeds to sniff the air intently. We walk over toward camp, toward the fox, but she seems undeterred. We stop about thirty feet away to await her next move. After another minute or so she stands up, suddenly, sticks her tail straight out so her body is like a long, sleek arrow, and trots off through the sedges to a small cluster of willows nearby. I suspect that there might be bird nestlings of some kind that she has heard, but there are none, it seems, and the place is still. The vixen sets up in the tall sedges next to the willows and just waits, patiently. A minute goes by. Absolute stillness. Then, suddenly, as quick as lightning, she darts for something, just like an arrow, deftly and professionally. There is a loud squeak, and a second later the proud vixen emerges from the bush with the kicking form of a plump arctic ground squirrel in her maw. She shakes it vigorously to snap its neck, then places in down in the sedges, glances over at us briefly, and sits down to look at her kill. The wind blows again and fluffs her coat, and I see the pink flash of her tongue as she cleans her mouth. She almost appears to be smiling. Then she picks up the dead squirrel in her mouth and trots right by us, right through the middle of our camp, and down the riverbed. Her kits are down there some-where, I presume, hidden in some secret den. Granny and I sit there dumbfounded. The presence of mammals pervades this place. And the presence of mammals that eat mammals.

Burwash Uplands

Burwash Creek

# X

We move camp up to Hoge Pass, on the alpine tundra, and spend several days exploring the surrounding country. The tundra here is different from the muskeg below because the plants that grow here are rooted in soil, not moss, and the organic layer that covers the earth is only a few inches thick instead of a few feet. The ground is noticeably firmer, and traveling, thankfully, is easier.

However thin the vegetation may be, it is no less fascinating. In moist areas, polar willow, netleaf willow, reindeer lichen, and mountain avens form an elaborate system of interlaced, prostrate woody stems and leaves. In dry areas, where vegetation is sparse, bulbous cushions of moss campion dot the landscape, and tufts of broad-leaved willow herb, or dwarf fireweed, garnish the edges of gravel flats. In wet areas, stands of cotton sedge hold their wispy, cottony seeds up to the wind. Heaths are conspicuously absent. The soil here derives from limestone and is not acidic enough for the heaths.

I am distracted from the plants by a scar of obviously disturbed soil in a gravel flat two hundred feet from my tent. It seems new to me, and I don't recall seeing it yesterday or the day before. As I wander toward it, I pick up the trail of a large grizzly. The tracks seem fresh, imprinted deeply into the soft, wet, sandy gravel, but it is difficult to tell exactly how old they are. They lead directly toward the scar in the hillside. There, the ground had been torn to pieces, revealing half a dozen dug-up ground squirrel dens, now silent and emptied, the dried fodder collected by the squirrels in piles where they left them. Grizzly tracks and claw marks surround the place.

We gather as a group in the kitchen area. Clad in raingear from head to toe, equipped with extra layers of warm clothes, food, drink, ice axes, and helmets, we prepare for a day of exploring the nearby Burwash Glacier. It is a small glacier in this place of immense ones, but huge glaciers can be difficult to take in, and the Burwash is reputedly well featured and relatively benign. After checking in, we set off over the hills, into the clouds and wind and rain.

We cross a series of long, lateral hills where yesterday I sat out and watched a herd of fifty Dall's sheep eat and rest and scratch their backs on overhanging mounds of sod. The hills run parallel to one another, each of them successively higher, separated by parallel drainages, each filled with gravel and rivulets of cold, clear water. We travel over this landscape for an hour, over a hill, into a drainage, over a bigger hill, into a drainage, over a bigger hill, toward the swirling, gray, low clouds that fill the Burwash Valley, spilling out over the surrounding landscape. As I look out over the rounded hills, periodically obscured by mist and intermittently washed with warm, glowing sunlight, I am overwhelmed by the feeling of so many other places in the circumpolar North. Scotland. Norway. Iceland. The tablelands of Mount Katahdin in Maine on a cloudy day. But then I remember the empty dens of ground squirrels and the grizzly tracks around my tent, and I am reminded of where I am.

We cross broad, scree-covered slopes at the tops of subtle drainages and climb onto a recent lateral moraine of the yet-unseen glacier. Loose flocks of pipits and rosy finches sprinkle the otherwise still landscape. The presence of the moraine means the ice must be near, as it is a fresh moraine, unsettled and unvegetated, a remnant of the Little Ice Age. We follow the moraine downslope, toward a huge tower of cracked, polished limestone, 150 feet tall and 30 feet wide, that juts out into the glacial valley. We find the rock bridge that connects the tower to the surrounding landscape and follow it out to the top of the pillar. It is like an immense battlement surrounded by a valley of mist, cracked in places where it has broken and fallen into the abyss that surrounds it. Cracks run even through the middle of its top, reminders that the tower might fall at any time. We sit, perched above the valley, and wait for the parting of the mist.

Within minutes, we notice small, conspicuous patches of blue open up over Burwash Creek below, and the rain that has not stopped since last evening begins to relent. The mist in the valley below shifts and swirls defiantly, then it, too, slowly begins to tear open and spin away in shreds. As the curtain rises, the diminished form of the snout of the

Burwash Glacier is revealed. It sprawls downvalley like a dry, white tongue, cracked, stiff, and shrunken, its leading edge indistinguishable from the debris that covers it. Concentric mounds of horseshoe-shaped moraines surround the ice, further evidence of the recent demise of the glacier. Judging by the size of the moraines, I estimate that the ice is presently well less than a tenth the volume that it was at the height of the Little Ice Age. Even the smallest, innermost moraine stands in heaps above the current level of the ice. I am reminded once again that we are losing it. The ice is leaving this world.

Downvalley, the milky outwash of the Burwash Glacier emerges from the debris-blackened terminus and crashes down through a steep, confused valley of crushed till and dead ice, some of which blocks the valley like a dam, but which has been tunneled through by the rushing water. Below the ice dam, the meltwater is constricted by a narrows of limestone several hundred feet high, and flows through a slot in the rock wall ten or twenty feet wide. Beyond the narrows, the meltwater spills out onto flat ground and splits into a dozen braids.

We make our way back up the outermost lateral moraine of the glacier, tiptoeing along its loose spine toward the glacial basin above. Where the moraine ends, we carefully traverse a precarious scree slope that is underlaid by ice, and we send tons of fist-sized rocks sliding downslope. Moving single file, we reach the edge of the ice and step out onto more stable terrain. Once we are all on the glacier, the traveling becomes easier. The surface of the ice is benign, perfectly textured for walking by a peppering of sand and pebbles. We ascend quickly to a plateau at the midriff of the glacier, admiring the curvacious incisions that runnels of meltwater have made in the ice. Eventually, the bare ice becomes covered with a two-inch layer of slush—the melting snow of the last two days of storm. Here, we stop for a rest and to make a plan.

We sit awhile amid the shifting light, tattered wisps of clouds swirling around the upper reaches of the glacier and obscuring the high peaks, the basin where we sit periodically bathed in warm sunlight, then cast in gray once again. Granny and a student, Sam Fox, made a reconnaissance in this direction yesterday and reported strange mammal activity on a ridge near the glacier, and I had hoped to find mammal signs. We peer out across the ice, but find none. Then, just as we are preparing to descend, up on a small bib of ice to the right of the head of the glacier, I spy tracks. Hundreds of them. Big animal tracks in the snow.

We practically run up the ice, hoping to catch at least a glimpse of an animal that has undoubtedly known of our presence for a long time. When we reach them, we find that the tracks are fresh, made today. Anything older would have been covered by last night's snow. It is instantly apparent that there were several animals here, at least two different adults and two young. The tracks are everywhere, up, down, and across the slope. The largest are as long as my hand and almost as wide, the forefeet like those of a wolf but with five toes, the hinds long and triangular, with disproportionately huge claws. In places, the animals have obviously dragged something big and heavy upslope, across the ice. Whatever it was, it was big enough to erase their tracks. I follow the drag marks upslope and find fresh droppings melted into the snow. Then I find hair, tan and brown, hollow, the hair of some kind of deer. Then pieces of old dried flesh. A trail of these scraps leads to a large mound of snow out of which various bones and body parts protrude, the half-buried, mutilated carcass of a caribou, a single rotting hoof the only clear indication that this macabre wreck of a body was once the very same animal we saw trotting across the muskeg just days before. Half a dozen such piles litter the snow. Hundreds of tracks surround the carcasses.

*Gulo gulo*, the wolverine, the most clandestine and elusive of all predators, second to none in ferocity and tenacity, with the temperament of a giant badger, the strength of a small bear, the guile of a weasel, the wanderlust of a wolf, an incomparable endurance in the face of cold, snowy, desolate conditions, and the reputation of a spirit-demon from the netherworld was here, in this place, and did this. And we are in its haunt.

Wolverines play a fearsome role in the lore of the north for a multitude of reasons. At twenty-five pounds, they have been known to pair up and take down six-hundred-pound caribou, even thousand-pound moose. Occasionally a wolverine will take down such a prey singlehandedly. But they are mostly active scavengers. If they can protect their haunches, they are capable of putting up such a ferocious display as to drive an entire pack of wolves from their kill. The wolves wait in view while the wolverine eats its fill, and only when the glutton stumbles off to its next meal do the wolves return to eat what is left. In other instances, when molested by grizzlies, wolverines have been known to roll onto their backs and disembowel their attackers. As scavengers, they wander far and wide in search of food. In fact, unless they are raising young, wolverines never stop moving. They are indefatigable, capable of covering hundreds of square miles in a day. They stay active year-round, through the bitterest of northern winters, stoking their internal fires whenever they can by stealing, scavenging, and killing huge quantities of meat. Countless tales of the north tell of the demon of the snows that comes to raid the meat caches, driving people to hunger, starvation, and sometimes death. More than an actual physical menace, over the centuries the wolverine has become a kind of phantom of the worst kind, one that is unpredictable, elusive, and comes in the night.

Other carnivores eat meat, but few eat it exclusively. Dogs, foxes, bears, and even cats vary their diets to some degree with plant matter. But the members of *mustelidae*, the weasel family, among which the wolverine is the largest, eat meat, meat, and more meat. The wolverines of the Burwash area have dragged at least two caribou carcasses up and onto the ice, quartered them, and refrigerated them by piling snow on top of them. They are incredibly intelligent, as it takes a fair amount of problem solving to figure out that only above the firn line, the line above which the glacier is covered with permanent snow, will they be able to bury their cache and keep it cold. They return hour after hour, day after day, even week after week to this cache to feed themselves and their young. Above us, a neat line of tracks trails off through the snow and over a steep talus slope to their den. It is their getaway line from the host of intruders that is upon them.

Caribou are common down on the muskeg, where their favorite foods abound. But the steep upper slopes of the alpine tundra are the realm of the Dall's sheep, and caribou are relatively scarce here. Could the wolverines have dragged the caribou carcasses all the way up from the muskeg, miles away and over three thousand feet below, or were the caribou tinkering around near the lower glacier, nibbling on scattered alpine plants? Had the caribou been killed by wolves and scavenged by the wolverines, or did the wolverines themselves take down such large prey? Questions tumble out of our mouths as we try to piece a story together. Speculations aside, we go over the facts: at least four wolverines were here today, and have been here in days past. Two of them appear to be adults, and two young. They have been feeding on the quartered carcasses of at least two different caribou. The carcasses have been dragged up the snow recently, and are buried in packed firn. The tracks lead off to the talus slope on the side of a nearby peak.

We wait for while up on the ridge, where the tracks leave the snow, in hopes that we might catch a glimpse of the wolverines. One student, Alexis Finley, arrived there in time to see two amorphous shapes moving among the rocks a few hundred feet away, but now they are gone. After some time, maybe half an hour, we decide we have harassed the animals long enough, and we descend back through the carnage and onto the glacier.

The indescribable feeling of having just witnessed something rare and incredible pervades the group as we walk down the bare ice. We are under the influence of some kind of ineffable spell of the wild, and there is a glint in everyone's eyes, a sense of sharpness, a sense of being in tune with something deep and primal. On either side of us, melt-water rushes down spiraling grooves in the ice, and they grow to miniature glacial slot canyons as the runnels converge and increase in volume. We toss pebbles into the slurry, then rocks, then head-size boulders, and watch as they career through the slots, even the largest of them suspended in the wild, fluid motion of the water. Then they fall into moulins, dark and deep, and rumble down into the unknown of the bowels of the ice, only to converge in hidden recesses and burst forth at the snout in a rushing river of meltwater. We move swiftly, without words, and follow the water down.

*Gulo gulo*

Ice cave

# XI

Andy Williams casually starts up his plane, an old '66 Courier built in Kansas City, with the easy manner of one who has been flying for decades. He says something in his washed-out Welsh accent, smiles, and laughs heartily. As he does, the cockpit is filled with the aroma of tobacco. I return the smile, but whatever he said was inaudible over the overwhelming whir of the engine. I settle into my seat as we roll down the dirt runway, kicking up a cloud of dust behind us. Andy reaches up to the ceiling and comically turns a large, elbow-shaped crank. He winks and smiles again as I realize that he is lowering the wingflaps. We are ascending. Within seconds, Kluane Lake is a blue void beneath us, and the gray-brown heavily silted braids of the Slims River fill the windows.

We follow the Slims River to its source, the downwasting snout of the Kaskawulsh Glacier. These days, the rate at which the lower portion of the glacier ablates is faster than the rate at which the upper portion accumulates; thus, the ice here shares the fate of nearly all the other glacial masses of the world, and is shrinking. Nevertheless, it is magnificent, and its scale is grand. From the demolition zone of the glacier's terminus, we fly along its length. It is like a huge white road of ice, bound on all sides by mountains streaked with medial moraines that appear as neat, parallel, black lines. Where the glacier curves around mountain buttresses, the lines curve, too, as if in formation, complementing the ice with graceful accents. The surface of the ice is bare this far down, this late in the summer. It appears smooth from four thousand feet up, subtly laced with an elaborate network of surface runnels. Only where the ice curves and bends is the surface complicated by tight rows of innumerable crevasses. The Kaskawulsh is one of scores of valley glaciers flowing off the main mass of the icefields, but at four miles wide and over forty miles long, it is huge in and of itself. Not only is it huge, it is also impressive, relatively accessible, and often photographed. I have seen its portrait in half a dozen books on glaciology.

As we continue up the glacial highway of the Kaskawulsh, past the tributaries of the South Arm and the Stairway Glacier, the mountains on either side of us grow higher and increasingly caked with ice. Their flanks are literally spilling over, ice tumbling in Slinky-like icefalls and neatly curving smaller valley glaciers, down to merge with the larger Kaskawulsh. But many of these tributary mountain glaciers are hanging high up on the mountain flanks, cut off from the master stream. These suggest a time when the Kaskawulsh was much more voluminous than it is today, its surface much higher, a time when all of the hanging glaciers were directly connected to the Kaskawulsh, rather than feeding it indirectly through icefall.

Forty miles up the Kaskawulsh, the surface of the ice abruptly and conspicuously turns white. We cross the firn line and the superficial features of the glacier become obscured in permanent snow. As we near the head of the Kaskawulsh, the tributary glaciers flow directly into the main mass, those remaining hanging glaciers on slopes too high and too steep ever to have been directly connected to the main body of ice. As we near the main icefields, the Kaskawulsh seems to melt into an endless expanse of white, and the view becomes Greenlandic, Antarctic, Pleistocene in scale, nothing but ice to the horizon in all directions, mountains poking out as mammoth islands of frozen stone. Celestial Mount Vancouver. Mysterious Mount Saint Elias. Crystalline Mount Logan. They are enormous, every one of them, standing eight, ten, and twelve thousand feet above the ice.

I have a map room idea of where we are, but with so much white below us I become lost. Each time my expression grows confused, Andy leans over, as if on command, and shouts the name of the feature I am looking at. The Logan Glacier. The Hubbard Glacier. Mount Queen Mary. Mount King George. But one feature I am certain of. As we fly west, the ice of the Seward Glacier spills through a pronounced gap in the mountains and sprawls out to the purple sea in a single fan-shaped lobe of ice fifty miles wide. It is the Malaspina Glacier, the largest Piedmont Glacier in the world.

"So, ah, we must be over Alaska now, huh?" I yell over the roar of the propeller.

"Ah, yes, that's somewhere around here," Andy replies with a sly grin.

"What does the U.S. think about us flying over the border?" I ask. Andy's grin becomes a chuckle.

"Well, now, they don't like that too much."

But Williams is the archbaron of these skies, and therefore does what he wants. While other outfits offer flights over the icefields, Williams is the only pilot who can actually land. In fact, during spring and summer, his is the only plane in the Yukon Territory equipped with the skis necessary for glacier landings. His livelihood depends on this fact, and mountaineers and researchers from far and wide depend on him and his ski-equipped Courier (Williams always flies Couriers) to get them onto the ice. Thus, the Icefield Ranges, over his thirty years of flying here, have become a familiar to him as the road to work has for most modern North Americans. Of the half dozen phone calls I made inquiring about glacier pilots in the Icefield Ranges, every single one of them led me to Andy Williams.

We hover around the Icefields like a tiny white hummingbird, easily lost in the gargantuan scale of the landscape. Eventually we recross the international border somewhere over the Hubbard Glacier and follow it back toward the Kaskawulsh. From there, navigating is easy, and the glacial highway leads us back out toward the braided outwash plain of the Slims River. As we pass over the debris-covered terminus of the glacier, I study it closely. Tomorrow we will head up the Slims River on foot and make our way up onto the Kaskawulsh for a multiday exploration of the glacier. I am looking for an easy way onto the ice.

On the ground, the Slims River is like a mudflow, and there seems to be no difference between mud and silt. Mysteriously, unlike every other river I have seen in the north, there are no gravel bars here, not even one. Only silt bars. In fact, the Slims is the most silt-laden river in all of the Yukon. The river is called the Slims because of an unlucky horse that died in the quagmires, sinking ever deeper into the bottomless mud, ultimately to meet its end. All this silt means nutrients and alkalinity in the soil, making an ideal habitat for soapberry, which makes an ideal habitat for grizzlies. Thus the Slims, being the siltiest river in the Yukon, hosts the highest population of grizzlies in the territory, and as the Yukon has the highest concentration of interior grizzlies in the continent, we find ourselves smack-dab in the middle of what is probably the highest and densest grizzly population in the world.

Because of the quagmires, the riverbed is impassible, so we stick to the east bank of the river, where we thrash through willows and soapberry, singing and clapping noisily as we walk. In one hour we see more bear tracks than I have seen in my whole life. Big tracks, small tracks, old tracks, new tracks made so recently I can still make out the minute cracks in the toe pads. There are tracks in silt, tracks in sand, even tracks imprinted into aquatic vegetation. It becomes obvious to us that bears walk the Slims River corridor daily, even hourly, and it is not a question of *if* we will see them, not even a question of *when* we will see them, but rather a question of *what we will do* when we do see them. I sing with vigor.

The upper Slims River Valley is shrouded in tattered clouds swept inland by westerlies that blew throughout the night. Clouds seem to pour over the mountains and ridges, cloaking the landscape in a forboding mystique. In the gray and rain and cold wind, I am reminded of the Pleistocene. It occurs to me, as we approach the snout of the still-to-this-day massive yet severely degraded Kaskawulsh Glacier, that the scene before me is perhaps as close as I will ever get to experiencing what the landscape was like in the last days of the Ice Age.

The flat silt plains of the Slims River, rich in carbonate minerals contained in recently crushed and powdered stone and wind-blown loess from the glacial outwash, hosts a sparse but nutrient-rich vegetation of grasses, sedges, and sagebrush, a small suggestion of the once-vast Arctic steppe that may have once covered much of the North during the retreat of the Pleisticene glaciers, when such nutrient-rich soils covered the ground. These great northern steppes may have supported vast herds of large grazing animals such as the infamous wooly mammoth. Most of the remains of the now-extinct Pleistocene megafauna have been excavated from old riverbeds, and the number of these finds suggests that there may have been large populations of these mammals throughout the North. More skeptical researchers argue that although the abundance of these finds suggests a lot of animals, lots of animals along the riverbeds does not necessarily mean lots of animals out on the range. I think the skeptics are onto something. Common sense suggests that animals congregate around water sources for a plethora of reasons, and empirical evidence supports this idea. In our limited time in the North, we have come across a score of different sets of bones and antlers, and all but one of them has been in a riverbed. We have found more tracks along watercourses than anywhere else. In range-land throughout the semiarid North American West, cattle tend to hang out around water sources. It should come as no surprise that similar demographics held true among the animals of the Pleis-tocene. It is not likely that mammoths and their herbivorous and carnivorous associates were as dense across the whole of the North as they were along watercourses. But skepticism is often taken too far. Throughout the Pleistocene, and especially during the period from fourteen to nine thousand years ago, during the great melt, watercourses and their associate outwash was everywhere. Imagine the Slims, with its soapberry flats, wide swaths of steppe, and high population of grizzlies, multiplied by a million. The sum includes a whole lot of big, furry animals. Steppe probably covered the ground, and herds of megafauna, though perhaps not quite as large as some might suggest, were strewn across the land.

The Little Ice Age moraine of the Kaskawulsh Glacier forms an enormous horseshoe-shaped rampart of piles and piles of pushed-up, ice-cored, unsorted debris that lies about half a mile downvalley from the snout of today. The moraine forms a dam behind which much of the meltwater from the glacier pools sits, dropping its silt among silent, debris-covered chinks of dead ice to form a complex system of still pools, saturated silt flats, and quagmires. If I could sum up the landscape in one word it would be this: *crushed*. We follow a winding route through the *crushed*, careful to test silt flats before venturing out on them, staying on sinuous beds of raised debris whenever possible. Soon we find ourselves going up, onto heaps of ice-cored debris. We travel over a dozen such debris ridges as we make our way to bare ice. This uneventful and vague transition marks our long-anticipated emer-gence onto the ice of the Kaskawulsh Glacier.

Stepping onto the bare ice feels good, and almost immediately the feeling of foreboding that had begun to overcome the group lifts and dissipates. *This* is ice. *Big* ice. It is fantastic, cracked, groaning, blue, gray, white, gurgling with meltwater above and within, the last of the Pleistocene beneath our feet.

We make camp on a fifty-foot-wide strip of ice bounded on each side by a hundred-foot-long crevasse. The ice groans and pops like gunfire. There are bones and bear scat around camp. The weather makes us cower. There is not another human being for a hundred miles to the west and twenty miles in any other direction. I am so glad to have found these precious last wild things.

Kaskawulsh Glacier                                    Icefall

# XII

We spend five days on the ice. It is the last five days of our expedition; our final stint in the mountains of the North American Cordillera. It rains the whole time, and the mountains around us become mere figments of our imagination. The Kaskawulsh is transformed into a floating bed of ice, like some Olympian platform suspended aloft in a world of cloud. A katabatic wind blows off the main icefield, steadily down the glacier, and the effect is something between a swamp cooler and a refrigerator. We move camp several miles up the glacier, and from there we scout still farther up the ice in search of features: deep, jumbled crevasses, rounded pockmarks filled with jet-black pebbles and still sapphire water, runnels the size of rivers, the icy blue holes of moulins big enough to swallow trucks, where rivers disappear into the unimaginable interior of the glacier. Each day the surface of the ice melts down several inches, as evidenced by the slushy platforms that form beneath the shade of our tents. It doesn't take much to figure out how much water so much melt amounts to. The glacier is four miles wide and forty miles long before the firn line, which makes sixteen hundred square miles. Every day, a four-inch thick sheet of water that big sheds off the Kaskawulsh, swelling the Slims River to its banks.

For our last night on the glacier we have found an extinct moulin, a round, vertical cavern that descends directly into the ice, that we intend on venturing into. It was formed by the surficial river that runs through the middle of our camp, which has since found a more direct route into the glacier (via a new and improved moulin) and left the old moulin dry and empty. We set screws into the ice at the lip of the moulin, and one by one rappel down into the cold interior of the Kaskawulsh.

I drop thirty feet into the moulin and brake to have a look around. I am surrounded by turquoise and royal blue ice that is as hard and as clean as glass. I reach out to touch the smooth surface and find it covered with the thinnest possible veneer of water slipping perpetually down its vertical surface. Above me, the upper thirty feet of the hole forms a perfect, five-foot-wide tunnel into the glacier. Just a few feet below the surface, the ice turns smooth, more like water ice than glacial ice, fluvially sculpted like marble or long-worked granite in a streambed. It looks almost like blown glass in places, different from anything else I have seen in a glacier because of its pure roundness.

I continue down five more feet to an uneven ice ledge. A horizontal passage continues to my left, where the water that once flowed through here made a right angle turn for a while before plummeting to greater depths. I stuff myself into the passageway and scoot through the wet, polished interior. The passage continues for ten or twelve feet and drops abruptly into a dark chasm, an interior crevasse of some kind that has been further sculpted by water. Down deep there, where the brilliant blue light is overcome by blackness, I can just barely make out the sound of water rushing through the bowels of the ice, bound for the outwash plains of the Slims River Valley. The roaring mixes with the clinking of my crampons as I work my way back through the passageway to the ice ledge at the bottom of the hole. Bright light opens up above me and again the ice is illuminated in brilliant hues of blue. I stand there just breathing for a while, then call to the crew to haul me up. I emerge wide-eyed, back in the world of light, back in the world of the living.

In the morning we awaken to smoke. The summer fires raging across the Yukon have cast a smokescreen over the Icefield Ranges that replaces the gray of storms with a pale, brown haze. It is so thick that we cannot tell where the sun is in the sky, though we know from the light that it is out there somewhere. Our throats are sore, eyes bleary, and we wonder if it will rain ash on us today.

Although we do not speak of it aloud, it seems strange to all of us that our time in the mountains is about to come to an end, that we can go so quickly from inside the crystal clear blue heart of a glacier to the ambiguities of modern American culture, that seventy days and nights along the North American Cordillera are about to be behind us, and what lies ahead is shrouded as physically as it is emotionally in an impenetrable veil of smoke.

I am thinking about the nature of exploration and the nature of the return. These thoughts always come into my mind at the end of a season outside. History has not always been kind to those intrepid individuals who have left their everyday lives and ventured out into the wider world. While some meet with the rich rewards of discovery, all must reckon with the profound and often abrupt psychological shift that occurs when it is time to return to wherever they came from. Upon return to society, the simple, purposeful roles we play as explorers quickly become complex and often seemingly meaningless. The senses that we have so carefully tuned are suddenly assaulted with a barrage of modern stimuli. The bodies that we have worked so hard to hone are swiftly made soft and listless under the sedative influence of modern comforts. Some, such as Thomas Mallory, never return, but rather meet their worldly ends in elemental nature, swallowed whole by the world around them. Others return to find the life they left behind just as meaningless as it was when they left. Of these, some, such as Reinhold Messner, are called back to explore again and again, as if cursed (or blessed) by an incessant need to always be exploring, always going forth on some expedition of discovery. Others, such as Meriwether Lewis of the Lewis and Clark expedition and Halmar Johanson of the Norwegian expedition to the South Pole, lose themselves in society, and despair. A few explorers, such as Edmund Hillary, manage to find meaning in everyday life, and happily live to grow old. But no explorer, no matter how short or long the journey, ever returns home the same.

It occurs to me, as I look out across the ice, that we are like droplets of water. We leave our source and travel far, changing, becoming air, becoming water again, falling down over the earth, journeying through these mountains and glaciers and lakes and rivers, perhaps for years, even eons, but always returning to the place from which we came, the sea, itself so enormous and unfathomable it is overwhelming. Then, some time, we begin the cycle again; the same theme, but a different story. Over and over again.

We pack slowly, methodically, quietly, and set out across the ice. As we walk, the orange orb of the sun emerges from the haze and hovers before us, over the valley of the Slims. We walk toward it, out of the mountains, into the smokescreen, following the course of a hundred gurgling, cold runnels as they leave the ice to join the river.

Surface melt                    Departure

# Bibliography

Ahrens, C. Donald. 1991. *Meteorology Today: An Introduction to Weather, Climate, and the Environment*. St. Paul: West Publishing Co.

Alt, David and Donald W. Hyndman. 1975. *Roadside Geology of California*. Missoula, MT: Mountain Press Publishing Co.

Armington, Stan. 2001. *Trekking in the Himalayas*. Victoria, Australia: Lonely Planet Publications.

Arno, Stephen. *Discovering Sierra Trees*. 1973. Yosemite Association, Sequoia Natural History Association: National Park Service, U. S. Department of the Interior.

———. 1984. *Timberline: Arctic and Alpine Forest Frontiers*. Seattle: The Mountaineers.

Barbour, Michael G. and William Dwight Billings, eds. 1988. *North American Terrestrial Vegetation*. Cambridge: Cambridge University Press.

Barbour, Machael G. and Jack Major, eds. 1990. *Terrestrial Vegetation of California* Sacramento: California. Native Plant Society.

Barry, Roger G. and Jack D. Ives. 1974. Introduction, pp. 1–11. In Ives, J. D. And R. G. Barry, eds. *Arctic and Alpine Environments*. London: Methuen.

Benn, Douglas I. and David J. A. Evans. 1998. *Glaciers and Glaciation*. London: Arnold.

Bernbaum, Edwin. 1990. *Sacred Mountains of the World*. San Francisco: Sierra Club Books.

Billings, W. D. and L. C. Bliss. 1959. An alpine snowbank environment and its effects on vegetation, plant development, and productivity. *Ecology* 40:38897.

Billings, W. D. and H. A. Mooney. 1968. The ecology of arctic and alpine plants. *Biological. Review* 43:481–529.

Billings, W. D. 1973. Arctic and alpine vegetation—similarities, differences, and susceptibility to disturbance. *Bioscience* 23:697–704.

———. 1974. Adaptations and origins of alpine plants. *Arctic and Alpine Research* 6 (2): 129–142.

———. 1978. Alpine phytogeography across the Great Basin. *Great Basin Naturalist Memoirs* 2:105–117.

———. 1988. Alpine vegetation. In Barbour, M. G. and W. D. Billings. *North American Terrestrial Vegetation*. Cambridge: Cambridge University Press.

Bone, Robert M. 2003. *The Geography of the Canadian North: Issues and Challenges.* Ontario: Oxford University Press.

Bonington, Chris. 1992. *The Climbers: A History of Mountaineering.* London: BBC Books.

Brower, Barbara. 1991. *Sherpa of Khumbu: People, Livestock, and Landscape.* Delhi: Oxford University Press.

Bundesamt fur Landestopographie. 2002. *Jungfrau Region* 1:25,000. Wabern: Bundsamt fur Landestopographie.

Bundesamt fur Landestopographie. 1998. *Zermatt Gornergrat* 1:25,000. Wabern: Bundsamt fur Landestopographie.

Burt, William Henry, and Richard Phillip Grossenheider. 1952. *A Field Guide to the Mammals.* Boston: Houghton Mifflin.

Canaday, B. B. and R. W. Fonda. 1974. The influence of subalpine snowbanks on vegetation pattern, production, and phenology. *Bulletin of the Torrey Botanical Club* 101 (6): 340–350.

Chabot, Brian F. and W. D. Billings. 1972. Origins and ecology of the Sierran alpine flora and vegetation. *Ecological Monographs* 42:163–199.

Chase, Clement G. and Terry C. Wallace. 1986. Uplift of the Sierra Nevada of California. *Geology* 14:730–733.

Christianson, M. N. 1966. Late Cenozoic crustal movements in the Sierra Nevada of California. *Geological Society of America Bulletin* 77: 163–182.

Clark, Douglas H. and Malcolm M. Clark. 1995. New evidence of late-Wisconsin deglaciation in the Sierra Nevada, California refutes the Hilgard glaciation. *Geological Society of American Abstracts with Programs, Cordilleran Section* 27:10.

Clark, Malcolm M. 1976. Evidence for rapid destruction of latest Pleistocene glaciers of the Sierra Nevada, California. *Geological Society of America Abstracts with Programs, Cordilleran Section.* 361–362.

Clyde, Norman and Wynne Benti (ed.). 1997. *Close Ups of the High Sierra.* Bishop: Spotted Dog.

Crough, S. Thomas and George A. Thompson. 1977. Upper mantle origin of Sierra Nevada uplift. *Geology* 5:396–399.

Department of Energy, Mines and Resources Canada. 1983. *Duke River* 1:50,000. Ottawa: Department of Energy, Mines and Resources.

Department of Energy, Mines and Resources Canada. 1987. *Slims River* 1:50,000. Ottawa: Department of Energy, Mines and Resources.

Diamond, Jared. 1999. *Guns, Germs, and Steel: the Fates of Human Societies.* New York: W. W. Norton and Company.

Dott, Robert H. and Donald R. Prothero. 1994. *Evolution of the Earth.* New York: McGraw–Hill.

DuFresne, Jim. 1998. *Tramping in New Zealand.* Victoria: Lonely Planet.

Duffy, Kevin. 1996. *Who were the Celts?* New York: Barnes and Noble Books.

Ellis, Reuben. 2001. *Vertical Margins: Mountaineering and the Landscapes of Neoimperialism.* Madison: University of Wisconsin Press.

Evans, Stephen G. and John J. Clague. 1994. Recent climatic change and catastrophic geomorphic processes in mountain environments. *Geomorphology* 10:107–128.

Fagan, Brian F. 2000. *The Little Ice Age: How Climate Made History 1300–1850*. New York: Basic Books.

Farquhar, Francis P. 1965. *History of the Sierra Nevada*. Berkeley: University of California Press.

Fiero, Bill. 1986. *Geology of the Great Basin*. Reno: University of Nevada Press.

Fowler, Brenda. 2000. *Iceman: Uncovering the Life and Times of a Prehistoric Man Found in an Alpine Glacier*. New York: Random House.

Gerrard, A. J. 1990. *Mountain Environments*. Cambridge: M. I. T. Press.

Gilbert, F. S. 1980. The equilibrium theory of island biogeography: fact or fiction? *Journal of Biogeography* 7: 209–235.

Gillespie, Alan R. 1991a. Quaternary subsidence of Owens Valley, California. In C. A. Hall, Jr., V. Doyle-Jones, and B. Widawski eds. *Natural History of Eastern California and High Altitude Research*, vol. 3. Los Angeles: University of California Press.

Gilligan, David. 2000. *The Secret Sierra: The Alpine World Above the Trees*. Bishop: Spotted Dog.

Gilligan, David. 2004. *The Mountains of California: Change and Evolution in Muir's Ranges of Light* (unpublished).

Grimmett, Richard, Carol Inskipp, and Tim Inskipp. 2000. *Birds of Nepal*. New Delhi: Prakash.

Hambrey, Michael and Jurg Alean. 1992. *Glaciers*. Cambridge: Cambridge University Press.

Harrison, Tom. 2000. *Mt. Whitney High Country Trail Map*. San Rafael, CA: Tom Harrison Maps.

Hartman, William K. and Ron Miller. 1991. *The History of Earth*. New York: Workman.

Heizer, Robert F. 1980. *The Natural World of the California Indians*. Berkeley: University of California Press.

Henry, J. David. 2002. *Canada's Boreal Forest*. Washington D.C.: Smithsonian.

Herm, Gerard. 1976. *The Celts*. New York: St. Martin's.

Hickman, James C. ed. 1993. *The Jepson Manual: Higher Plants of California*. Berkeley: University of California Press.

Hulten, Eric. 1968. *Flora of Alaska and Neighboring Territories*. Stanford: Stanford University Press.

Hunt, Sir John. 1954. *The Conquest of Everest*. New York: E. P. Dutton & Company.

Jeffries, Margaret. 1986. *Mount Everest National Park: Sagarmatha Mother of the Universe*. Seattle: The Mountaineers.

Johnson, Derek, Linda Kershaw, Andy MacKinnon, and Jim Pojar. 1995. *Plants of the Western Boreal Forest and Aspen Parkland*. Edmonton: Lone Pine.

Kennelly, Patrick J. and Clement G. Chase. 1989. Flexure and isostatic residual gravity of the Sierra Nevada. *Journal of Geophysical Research* 94:1759–64.

Krakaur, Jon. 1997. *Into Thin Air*. New York: Villard.

Kornfield, Jack ed. 1993. *Teachings of the Buddha*. Boston: Shambhala.

McNally, Rand and Company. 1989. *Deluxe Illustrated Atlas of the World.* Chicago: Rand McNally and Company.

Land Information New Zealand. 1996. *Aspiring* 1:50,000. Upper Hutt, NZ: Land Information New Zealand.

Land Information New Zealand. 2000. *Earnslaw* 1:50,000. Upper Hutt, NZ: Land Information New Zealand.

Land Information New Zealand. 1996. *Mount Cook* 1:50,000. Upper Hutt, NZ: Land Information New Zealand.

Land Information New Zealand. 2000. *Otira* 1:50,000. Upper Hutt, NZ: Land Information New Zealand.

Lindenmayer, Clem. 2001. *Walking in Switzerland.* Melbourne: Lonely Planet.

MacArthur, R. H. and E. O. Wilson. 1963. An Equilibrium Theory of Insular Biogeography. *Evolution* 17: 373–387.

MacArthur, R. H. and E. O. Wilson. 1967. *The Theory of Island Biogeography.* Princeton: Princeton University Press.

MacFarlane, Robert. 2003. *Mountains of the Mind: How Desolate and Forbidding Heights Were Transformed into Experiences of Indomitable Spirit.* New York: Pantheon.

MacMahon, James A. and Douglas C. Andersen. 1982. Subalpine forests: a world perspective with emphasis on western North America. *Progress in Physical Geography* 6: 368–425.

Magoun, Francis P. Jr., and Alexander H. Krappe. *The Grimms' German Folk Tales.* Carbondale: Southern Illinois University Press.

Major, J. and S. A. Bramberg. 1967a. Comparison of some North American and Eurasian alpine ecosystems. 89–118. In H. E. Wright and W. H. Osburn, eds. *Arctic and Alpine Environments.* Bloomington: Indiana University Press.

Major, J. and S. A. Bramberg. 1967b. Some cordilleran plants disjunct in the Sierra Nevada of California and their bearing on Pleistocene ecological conditions. 171–188. In H. E. Wright and W. H. Osburn eds. *Arctic and Alpine Environments.* Bloomington: Indiana University Press.

Major, J. and D. W. Taylor. 1977. Alpine. 601–675. In M. G. Barbour and J. Major eds. *Terrestrial Vegetation of California.* New York: Wiley.

Matthews, Daniel. 1994. *Cascade-Olympic Natural History.* Portland: Audubon Society of Portland.

Matthiessen, Peter. 1978. *The Snow Leopard.* New York: Viking.

McNally, Rand and Company. 1989. *Deluxe Illustrated Atlas of the World.* Chicago: Rand McNally and Company.

McSaveney, Eileen. 1996. *New Zealand Adrift.* Lower Hutt, NZ: Institute of Geological & Nuclear Science Limited.

Messner, Reinhold. 1989. *The Crystal Horizon: Everest—The First Solo Ascent.* Seattle: The Mountaineers.

Messner, Reinhold. 1999. *All Fourteen 8,000ers.* Seattle: The Mountaineers.

Molnar, Peter and Phillip England. 1990. Late Cenozoic uplift of mountain ranges and global climate change: chicken or egg? *Nature* 346:29–34.

Monasterski, R. 1993. Here comes the sun-climate connection. *Science News* 143: 148.

Moon, Geoff. 2000. *Common Birds in New Zealand 2: Mountain, Forest, and Shore Birds.* Auckland: Reed.

Mowatt, Farley. 2000. *The Farfarers: Before the Norse.* South Royalton, Vermont: Steerforth.

Muir, John. 1894. *The Mountains of California.* New York: Century.

Munz, Philip A. and David D. Keck. 1949. California plant communities. *Aliso* 2: 87–105.

———. 1949. California plant communities supplement. *Aliso* 2:199–202.

———. 1959. *A California Flora.* Berkeley University of California Press.

Nash, Roderick. 1967. *Wilderness and the American Mind.* New Haven: Yale University Press.

Ortner, Sherry B. 1989. *High Religion: A Cultural and Political History of Sherpa Buddhism.* Princeton: Princeton University Press.

Peat, Neville. 1994. *Land Aspiring: The Story of Mount Aspiring National Park.* Dunedin: Craig Potton.

Pielou, E. C. 1979. *Biogeography.* Halifax: Dalhousie University Press.

Price, Larry W. 1981. *Mountains and Man: a Study of Process and Environment.* Berkeley: University of California Press.

Pielou, E. C. 1988. *The World of Northern Evergreens.* Ithaca: Comstock.

Pielou, E. C. 1991. *After the Ice Age: the Return of Life to Glaciated North America.* Chicago: University of Chicago Press.

Pielou, E. C. 1994. *A Naturalist's Guide to the Arctic.* Chicago: University of Chicago Press.

Pojar, Jim and Andy MacKinnon Eds. 1994. *Plants of the Pacific Northwest Coast: Washington, Oregon, British Columbia & Alaska.* Vancouver: Lone Pine.

Quammen, David. 1996. *Song of the Dodo.* New York: Touchstone.

Razetti, Steve. 2000. *Trekking and Climbing in Nepal.* London: New Holland.

Reynolds, Kev. 2000. *Walking in the Alps.* New York: Interlink.

Roberts, J. M. 1993. *History of the World.* New York: Oxford University Press.

Salmon, John T. 1992. *A Field Guide to the Alpine Plants of New Zealand.* Auckland: Random House New Zealand.

Schaffer, Jeffrey P. 1997. *The Geomorphic Evolution of the Yosemite Valley and Sierra Nevada Landscapes.* Berkeley: Wilderness.

Sharsmith, C. W. 1940. *A Contribution to the History of the Alpine Flora of the Sierra Nevada.* Ph.D. dissertation, University of California Berkeley.

Siebeniecher, A. 1999. *Khumbu Himal* 1:50,000.

Simberloff, D. S. 1983. When is an island community in equilibrium? *Science* 220: 1275–77.

Skinner, Brian J. and Stephen C. Porter. 1987. *Physical Geology.* New York: John Wiley and Sons.

Small, Eric E. and Robert S. Anderson. 1995. Geomorphically driven Late Cenozoic rock uplift in the Sierra Nevada, California. *Science* 270:277–80.

Smith, Huston. 1958. *The Religions of Man*. New York: Harper and Row.

Starr, Cecie. 1991. *Biology: Concepts and Applications*. Belmont, CA: Wadsworth, Inc.

Stebbins, G. Ledyard and Jack Major. 1965. Endemism and speciation in the California flora. *Ecological Monographs* 1:1–35.

Stevens, George C. and John F. Fox. 1991. The causes of treeline. *Annual Review of Ecology and Systematics* 22:177–191.

Swan, L. W. 1967. Alpine and aeolian regions of the world. 29–54. In H. E. Wright and W. H. Osburn eds. *Arctic and Alpine Environments*. Bloomington: Indiana University Press.

Tabor, Rowland and Ralph Haugerud. 2002. *Geology of the North Cascades: A Mountain Mosaic*. Seattle: The Mountaineers.

Thoreau, Henry David. 1862. *Walking*. Old Saybrook: Applewood.

Tomlin, E. W. F. 1963. *The Western Philosophers*. New York: Harper and Row.

Trelawny, John G. 2003. *Wildflowers of the Yukon, Alaska, and Northwestern Canada*. Madeira Park, B.C.: Harbour.

Troll, Carl. 1973. The upper timberline in different climatic zones. *Arctic and Alpine Research* 5:A3–A18.

Unruh, J. R. 1991. The uplift of the Sierra Nevada and implications for Late Cenozoic epeirogeny in the Western Cordillera. *Geological Society of America Bulletin* 103:1395–404.

United States Geological Survey. 1974. *North Cascades National Park, Wash. And Lake Chelan and Ross Lake National Recreation Areas*, 1:100,000. Reston, VA: U.S.G.S.

Valdiya, K. S. 2001. *Himalaya: Emergence and Evolution*. Hyderabad: Universities.

Viereck, Leslie A. and Elbert L. Little, Jr. 2000. *Alaska Trees and Shrubs*. Fairbanks: University of Alaska Press.

Wardle, Peter. 1974. Alpine timberlines. 371–400. In Ives, J. D. and R. G. Barry eds. *Arctic and Alpine Environments*. London: Methuen.

Weeden, Norman. 1996. *A Sierra Nevada Flora*. Berkeley: Wilderness.

Went, F. W. 1948. Some parallels between desert and alpine flora in California. *Madrono* 9:241–249.

Williams, Hill. 2002. *The Restless Northwest: A Geological Story*. Pullman, Washington: Washington State University Press.

Whymper, Edward. 1986. *Scrambles Amongst the Alps*. Salt Lake City: Peregrine Smith.

Wu, John C. H., trans. 1990. *Tao The Ching*. Boston: Shambhala.

Young, Steven B. 1994. *To the Arctic: An Introduction to the Far Northern World*. New York: John Wiley and Sons.

Zurick, David and P. P. Karan. 1999. *Himalaya: Life on the Edge of the World*. Baltimore: Johns Hopkins.

Zwinger, Anne and Beatrice E. Willard. 1996. *Land Above the Trees: A Guide to American Alpine Tundra*. Boulder: Johnson.

# Index